CHEMISTRY OF
MARINE SEDIMENTS

edited by

T. F. Yen

Associate Professor of Chemical Engineering
Environmental Engineering Sciences
and Medicine (Biochemistry)
University of Southern California, Los Angeles

ANN ARBOR SCIENCE
PUBLISHERS INC
P.O. BOX 1425 • ANN ARBOR, MICH. 48106

Library of Congress Catalog Card No. 75-22900
ISBN   0-250-40103-7

PREFACE

The marine environment accounts for more than 70% of the global surface. Alone, the comparatively shallow regions of the continental shelf comprise an area about half that of the earth's lowlands, where most humans live. Consequently, the impact of marine sediments on energy, mineral resources, and food is exceedingly important for the near future.

Although there have been numerous marine and coastal studies, the chemical aspects of marine sediments have long been neglected. In this volume we first review the potential of marine sediments in energy development, for example, the genesis of petroleum. This is followed by a description of the chemical changes of fossil organic remains. The environmental effects of pollutants in sediments, such as heavy metals and pesticides, are then discussed, followed by a description of the interactions of sediments and sea water on organometallic pollutants. Last, a tool for investigating marine sediments is discussed.

T. F. Yen
Los Angeles, California

iii

ACKNOWLEDGMENTS

I would like to thank the American Chemical Society for having the foresight to sponsor a symposium under this title (many of the papers in this volume were presented at that symposium), all the authors for their patience in reworking and updating their papers, and Barbara James and May Louie for their valuable assistance.

CONTENTS

CHEMICAL ASPECTS OF MARINE SEDIMENTS

T. F. Yen and James I. S. Tang

Departments of Chemical Engineering,
Environmental Engineering, and
Medicine (Biochemistry)
University of Southern California
Los Angeles, California 90007

INTRODUCTION

Most of the ocean floor is covered by organic and inorganic particulate deposits that are called oceanic or marine sediments. Marine sediments are caused by unconsolidated accumulations of particles brought to the ocean by rivers, glaciers, and winds, mixed with shells and skeletons of marine organisms.

Marine sediments, as a tool of historical record for the ocean basins and of oceanic processes, such as nutrient supply to organisms and the effect of climate changes on ocean circulation, have long been studied. Much of the present knowledge of physicochemical and biogeochemical processes taking place within the sea stems from studies of these sediments.[1-7]

The bulk of the inorganic sediment is formed by the interaction of the atmosphere and the hydrosphere on the crust of the earth. Various aspects of the sedimentation process include weathering, erosion, deposition and diagenesis. With the participation of the biosphere in the above interaction process, the marine sediment composition containing organic molecules and its effect on the environment become important. As long as there are life processes there will be organic matter in sediment.

During the past few years, concern over ecological impacts caused by man's activity has brought extensive studies on the marine sediment as a pollution indicator. Since the sediment in the receiving environment may be an important reservoir of trace contaminants such as heavy metals and chlorinated hydrocarbons, sediment chemistry gives a measure of water quality and potential pollutants.

Most trace elements and nutrients can exist in several forms that differ in toxicity and availability. Physicochemical and mass transport phenomena and redox reactions of various chemical contaminants in the ocean waters and in the sediments caused by the activities of man govern to a large extent the propagation and distribution of chemical species among various available and nonavailable forms.

It is meaningful to consider that perturbations of geochemical fluxes by the activities of man will not be buried in submarine sediments, but will propagate and distribute themselves through the various natural reservoirs and fluxes because of the cyclic nature of the processes. [5,8,9] This is somewhat academic with regard to large-scale deep sedimentation because 1.2×10^8 years are required for the redistribution (or for the recirculation of all the sediments), but it becomes important within the smaller subcycles that occur in the atmosphere, hydrosphere, biosphere, and shallow sediments.[5]

Organic matter in sediment usually originates from the following atmospheric and riverine introduction of pollutants: industrial and domestic wastes, agricultural and mining runoffs, accidental spillages, decompositional debris from marine organisms, especially those bioresistant, and metabolic end products from natural biota. Furthermore, there are numerous forms of intermediates derived from the interactions among various decomposed products of living organisms. These terrestrial stable organic molecules can be treated as end products of different levels and stages in the geochemical diagenic process of simple bioorganic molecules.[10]

On the average the organic contents in marine sediments are only minor in composition. The organic carbon contents range from 0.1% to 10%.[11] However, regardless of their quantities, organic molecules remain primarily the controlling factors in marine sediments and their environments. The following reasons describe the importance of the organic compounds present in marine sediments:

(a) Organic molecules in marine sediments possess reactive functional group sites. The inorganic cations such as heavy metals can be coordinated to these sites to form stable linkages. In this manner the marine sediments can take up metals. The exchange capacity of sediments depends on the organic component of the sediment.[12]

(b) The metal complexes and chelates thus formed from (a) could further coordinate inorganic anions such as sulfate, chloride and phosphate and could be easily attached or detached under redox conditions. In this manner the transport or the migration of nutrition-important anions such as phosphate is regulated in sediment, an essential for marine life.

(c) The inorganic components in sediments behave as chromatographic substrates. Upon contact, the **organic** molecules could be preferentially adsorbed, fractionated, precipitated, eluted or desorbed. In this fashion, simple inorganic and organic molecules (pollutants) introduced into the coastal and estuary waters could be adsorbed and released. Due to exchange reactions, heavy metal ions can also be liberated from sediment to oceanic waters. Accordingly, the marine sediment is, in this sense, a source for heavy metal ions.

There may be several other ways to classify the type of sediments depending on the specific area of study, for example, nonpolluted and polluted sediments based on the degree of pollution, or nearshore and offshore sediments depending on the distance of the location from the land. Classifications based on their origin is as follows:

(a) *Lithogenous Sediment:* This type of sediment is primarily derived from breakdown of silicate rocks and carbonate shelves on the continents during weathering and soil formation. Often, in the open ocean, volcanoes are locally important sources of lithogenous sediment.

(b) *Biogenous Sediment:* The most common biogenous sediments consist of calcite, which is the insoluble remains of bones, teeth, or shells of marine organisms. Often, the chemically similar mineral argonite is found in the sediment. However, argonite is less frequently preserved in deep ocean sediments because argonite is more soluble in sea water. Siliceous shells of diatoms, radiolaria, and silicoflagellates are also common biogenous constituents.

(c) *Hydrogenous Sediment:* It is formed by chemical reactions occurring in sea water or within the sediment. Examples of the most common hydrogenous particles are manganese or iron-manganese nodules. These elements, originally brought to the ocean by rivers and possibly from volcanic activity on the ocean bottom, apparently precipitate in the presence of the abundant dissolved oxygen typically found in deep ocean waters. Their precipitates form tiny particles that are swept along by currents until they come in contact with a surface.

(d) *Cosmogenous Sediment:* It is composed of particles and objects that fall to the earth from outer space. A much larger percentage do not survive as recognizable particles but fall as bits of cosmic dust, much of it thought to have been from meteorites. Cosmic particles, named to be "cosmic spherules," are magnetic and composed largely of iron or iron-rich minerals. They are about 200 to 300 microns in diameter. Such spherules also occur in glacier ice.

In discussing the energy resources of the continental margins, the bathymetric areas must be defined. The earth consists of two topographic surfaces—the general elevation of the continents and the great depths of the ocean basins. The boundaries between these two regions comprise the continental margin of the world. The gently sloping sea floor adjacent to the shoreline is known topographically as the continental shelf, and its outer edge is marked by an increase in slope, known as the shelf edge. The steeper surface beyond the shelf is the continental slope, and the more gentle surface beyond the slope is the continental rise, which extends down to the average depth of the ocean basins at about 5000 meters. The shelf, rise, and slope together make up the idealized continental margin (Figure 1.1).

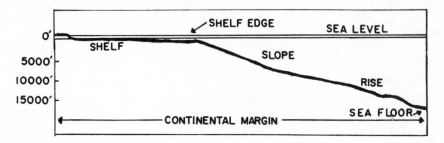

Figure 1.1 Schematics for continental margins.

In this review, we will outline the potential of various types of compounds within marine sediments and discuss the interrelationship, applications and limitations to the future resource exploration and management.

TRACE ELEMENTS

Authigenic mineral in pelagic sediments can provide a record of physicochemical and biological reactions operating and having operated at their site of deposition during the period of their formation. For instance, study of the layer-by-layer composition of manganese nodules might provide information on changes in the chemistry of the environment of deposition with time.[13-16] Similarly, study of radionuclides in marine barite, phillipsite, and phosphorite can provide information on the processes by which such elements are removed from sea water.[17-19]

Concentrations of hydrated manganese and iron oxides, in the form of nodules, are common features in the deep sea floor. The studies of the distribution of a number of trace elements in various marine environments showed that the average abundances of titanium, chromium, vanadium, gallium, rubidium and zirconium are similar in both near-shore and deep sea clays, but that the elements zinc, nickel, lead, cobalt, copper and manganese are more abundant in deep sea clays.[16] Barite ($BaSO_4$) is the most common mineral of barium and is widespread in deep sea sediments. Marine barites were found to have thorium and uranium concentrations of approximately 35 and 2.5 ppm, respectively. Other minor elements such as iron, chromium, zinc, copper, nickel, lead, manganese, cobalt and zirconium are generally present in marine barite in concentrations of between about 10 to 350 ppm.[1]

Contaminants and nutrients may exist in various forms and reside in different fractions of sediment, including the soluble form, exchangeable form, existence in carbonate mineral phase, easily reducible form, interactions with organic and sulfide fractions, association with a moderately reducible iron oxide or hydroxide, and presence in the lattice structure of clay and silicate minerals. Contaminants and nutrients in each fraction may be leached out under different environmental conditions.

Chen and Yen[12] and Yen[20] have studied the distribution of heavy metals and their mechanism of deposition in coastal waters along the California coastal lines (Figure 1.2). Marina del Rey is a recently constructed, man-made marina with two storm drain inlets. Los Angeles Harbor is a semienclosed basin with extensive industrial and marine usage. Santa Monica Bay is an open water with a sewer outfall. Long Beach Marina is a small craft harbor. The results of heavy metals in these stations are summarized in Table 1.1. In general, near-shore water samples contain no detectable mercury. Cadmium concentrations ranged from 2.5 ppb in Marina del Rey to over 10 ppb at the Los Angeles Harbor and Santa Monica Bay. The background level of cadmium in oceanic water is about 2-3 ppb. Lead contamination appears to be the most serious problem, ranging from 120 ppb at Marina del Rey and Santa Monica Bay to 260 ppb at the Los Angeles Harbor. The background level of lead in oceanic water is about 3-5 ppb. Fortunately such high concentrations of pollutants usually can settle down in several thousand yards toward the ocean.

The transport of trace elements between sediment and sea water may be a result of adsorption, desorption, precipitation, diffusion, chemical reaction, biological activity, and a combination of those phenomena. Although a complete solution of the exchange process between the sediments and the oceans will have to wait until the composition of and

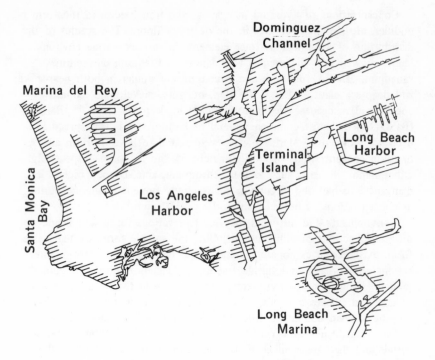

Figure 1.2   Location of Los Angeles Harbor.

Table 1.1   Heavy Metals in Water Samples (ppb)

		Pb	Cd	Hg
Marina del Rey		100-170	2.5-2.7	0.03
Los Angeles Harbor	Top	180-266	6.0-11.0	0.03
	Bottom	192-266	6.0-11.0	0.03
Long Beach Marina	Top	144-160	6.5-6.5	0.03
	Bottom	160	6.5-7.2	
Santa Monica Bay	Top	120-160	10.0-11.0	
	Bottom	160-240	10.0-11.0	0.03

gradients in pore waters are better known, progress is being made in sediment pore water studies [21-27] and rate of exchange across the sediment interface.[19,28-32]

Perry et al.[21] reported that throughout the sediment sequence, there is a steady depletion of magnesium and enrichment of cadmium in the pore waters. They explained the mechanisms as the following chemical reactions incorporated with other study results:[1,23]

$$Mg^{++} + 2\ CaCO_3 = CaMg(CO_3)_2 + Ca^{++}$$

(Dolomite)

or

$$6\ SiO_2 + 4\ CaCO_3 + 4\ Mg^{++} + 7\ H_2O$$

$$= 2\ MgSiO_6\ (OH) \cdot 6H_2O + 4\ CO_2 + 4\ Ca^{++}$$

The following diagenetic processes have been identified:[1,2]

(a) Dolomitization and recrystallization of calcium carbonate.
(b) Divitrification of silicates.
(c) Uptake of cations, especially magnesium and potassium, but also sodium, calcium and lithium in the form of authigenic silicate formation.
(d) Replacement of iron in clays by magnesium as a result of sulfate reduction.
(e) Ion-exchange reactions sensitive to temperature. In terms of relative temperature influence, the interstitial ions follow the order: boron, silicon, potassium, sodium, lithium (decreasing enrichment with increasing temperature), magnesium, strontium, calcium (depletion).

In near-shore areas, interstitial waters may be highly variable in composition. From the study of migration of trace metals (cadmium, chromium, copper, iron, mercury, manganese, nickel, lead and zinc) in the polluted near-shore sediment-sea water system, Lu[26] concluded that the migration is regulated mainly by the redox condition, especially the levels of dissolved oxygen and sulfide as the principal factor (Table 1.2).

Silica concentration in pore waters may be controlled to some extent by clay interactions in red muds (deep sea sediment with relatively no biogenous constituents) and by its solubility and upward diffusion in diatomaceous muds.[5] Veeh, et al.[19] reported that the organic-rich diatomaceous sediments containing authigenic phosphorite beneath areas of upwelling are the major sinks for uranium in the ocean. They also suggested that the mechanism for uranium enrichment in these sediments may involve the diffusion of hexavalent uranium (U^{6+}) from sea water

Table 1.2 Regulation of Migration of Sediments

Species	Redox Condition	Migration from Sediment to Interstitial Water
Fe, Mn	reducing	increase
Cd, Cu, Ni, Pb, Zn	oxidizing	increase
Cr, Hg	reducing & oxidizing	no specific change

into the sediment via pore water, followed by reduction to tetravalent uranium (U^{4+}) and its incorporation into carbonate fluoropatite growing diagenetically within the sediment.

The release of trace metals may be due primarily to desorption from iron, manganese, or clay minerals, and also from complex formation. Chloride, bicarbonate, and organocomplexes are the predominant species available as complexing agents. Feick et al.[33] found that chloride may cause the release of mercury to the aqueous phase. Formation of organometallic complexes may also cause the release of trace metals.[34-37]

The release of nutrients can be derived from the biodegradation of organic matter deposited on the sediment. It must be noted, however, that, unlike the sediments under shallow waters (or nearshore sediments that are prone to organic pollution), the organic matter in deep sea sediments is biologically more refractory.

The regeneration of nutrients in the sediments is suggested to be a two-step process: the first, which includes bacterial activity and physico-chemical reactions, determines the concentration of nutrients in the pore water; the second governs their release through the sediment-water interface.[38] Serruya et al.[38] reported that the exchangeable phosphorus seems to be the iron-bound fraction and its concentration in the pore water increases simultaneously with the reduction of sulfates and precipitation of iron sulfide, but its release is controlled by calcium phosphate equilibria.

Solubilization of $FePO_4$, $Ca_3(PO_4)_2$, $CaHPO_4$, and $Mg_3(PO_4)_2$ incorporated with the bacterial activity is attributed to the formation of organic acids, such as acetic, formic and lactic, that function as chelating agents, releasing free phosphate ions that can be utilized by algae.[39] Li et al.[40] reported that the rate of inorganic-P release into solution under anaerobic conditions is due mainly to the reduction of iron from the ferric to ferrous state. It has been mentioned that manganese has more predictive value than iron for determining the concentration of phosphorus in a given core of sediment; however, the iron content appears to be the dominant factor in the phosphorus sorptive and retentive capacity of sediment.

A mechanism has been suggested that ammonia, derived from nitrogen-containing organic matter, is converted to N_2 through a nitrate intermediate under conditions of low oxygen tension. Ammonia, however, is not converted to N in anoxic sediments that are undergoing sulfate reduction even though N_2 is the thermodynamically stable species.[41] Austin et al.[42] and Graetz et al.[43] found that the concentration of NH_3-N increased rapidly by nitrification. However, during the anoxic conditions, NH_3-N is released to the water at a relatively constant rate.[43] Vanderborght et al.[44] found that a maximum in nitrate concentration commonly appears at a few centimeters depth where sediments are sandy and poor in organic matter, while in muddy and organic-rich sediments, nitrate is lower in interstitial water than in the overlying water and decreases rapidly with depth.

HUMUS SUBSTANCES

It is a characteristic feature of the oceanic or sea water that it contains organic matter. The oceans and seas contain 2.6×10^{12} tons of organic matter, which is approximately equal to the world's resources of coal or peat.[45]

The predominant form in which organic matter is encountered in natural waters is as dissolved matter or water humus, which is profoundly transformed organic matter similar in properties to humus of terrigenous origin. The surface activity of humus substances is high and they readily form stable organo-mineral compounds with clay particles. Khan[46] stated that 0.53-3.77% of the organic carbon of humus substances is adsorbed from solution by various clay minerals, and that 24-76% passes into a firmly bonded and insoluble state. According to Bordovskiy,[47] a considerable part of the organic matter in recent marine sediments and sedimentary rocks is also in a firmly bonded and insoluble state.

Humic acids occur widely in various natural accumulations of organic matter, including soils, peats, brown coals, many fossil sedimentary rocks, and the range of recent subaqual deposits. The relative content of humic acids, or the humic factor, varies between 16.8 and 39.8% in Bering Sea sediments, and averages 26%.[47] Figures 1.3 and 1.4 show that approximately a quarter of the organic matter in the sediments consists of humic acids. The mean indices of the humic factor vary less widely by types of sediments; from 18.6% for medium grained sands to 29.8% for silt-clay muds. Tables 1.3 and 1.4 are the chemical composition of humic acids of recent Bering Sea and other sediments.

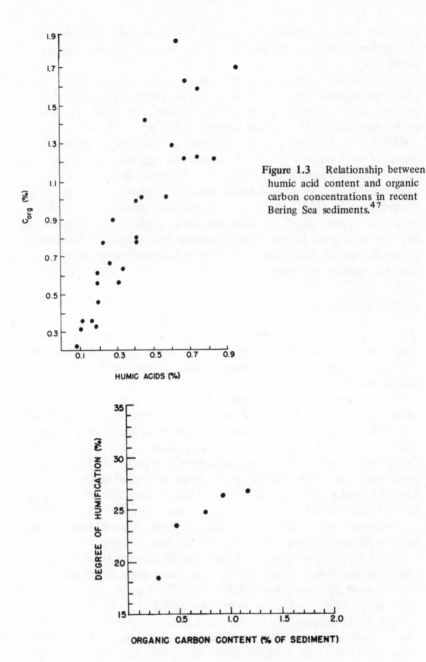

Figure 1.3   Relationship between humic acid content and organic carbon concentrations in recent Bering Sea sediments.[47]

Figure 1.4   Relationship between degree of humification of organic matter and total organic carbon content in Bering Sea sediments.[47]

Table 1.3 Recent Bering Sea Sediments

Type of Sediment	Depth (m)	C	H	C/H	N	C/N	O + S
Silt-clay mud	3451	54.75	6.30	8.7	2.98	18.3	35.97
Fine sand	508	53.93	6.10	8.8	3.08	17.5	36.89
Coarse silt	82	52.25	6.93	7.5	3.84	13.6	36.98
Coarse silt	67	49.23	7.16	6.8	2.80	17.6	40.81

Table 1.4 Chemical Composition of Humic Acids in Certain Recent Subaqual Sediments

Area	C	H	C/H	N
Bering Sea sediments	54.11	6.48	8.3	3.0
Sediments from Taman Peninsula reservoirs	55.66	6.32	8.81	5.55
Fore-delta sediments from the Caspian	59.17	4.85	12.2	3.82
Fresh-water sapropel	57.18	6.01	9.51	4.70

Young et al.[48] suggest that the precursor of humic substances are the melanoidins, or the browning products. They are formed from the hydrolyzed or decomposition product of cellulose and proteins. Simple α-amino acids and reducible pentoses can form melanoidin at very gentle conditions (pH = 7.6, room temperature) in aqueous systems. The relation between melanoidins and coals can be depicted in Figure 1.4.

The study of Degens et al.[49] gave the conclusion that humic acid is the dominant organic fraction in the surface sediments and accounts for about 30-60% of the total organic matter. Significant quantities of phenols and amino acids are associated with the humic acids, mainly by covalent linkage and, to a lesser extent, by ionic bonding or by chemisorption. The most convenient source for humic acids is lignins formed on the continents. The phenols attached to humic acids are identical to those found in humic acids of soils. Carbon isotope data are in support of the continental origin of the humic fraction. The overall geochemical spectrum suggests that most of the organic detritus supplied by the sea or generated *in situ* by burrowing animals and bacteria is biologically eliminated during the early stage of diagenesis. Organic nutrients, by some means attached to detrital clay minerals or terrestrial humic acids, are more stable and less effectively removed from the strata. Because of their abundance, humic acids are regarded as the dominant source material for kerogen in ancient sediments. The reaction of phenols with ammonia and their polycondensations can be depicted in Figures 1.5 and 1.6.

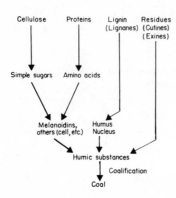

Figure 1.5   Relation of melanoidins to coal as derived from cellulose and protein.

Ishiwatari[54,55] gives examples of the hypothetical structure of the hydrocarbon corresponding to humic acid and asphaltic hydrocarbon in Figures 1.7-1.9. According to Ishiwatari, there may be three pathways of transformation of sedimentary humic acid into graphite, as shown below:

1. Sed. Humic Acid $\xrightarrow[-N_2]{-H_2O}$ "Soil Humic Acid" $\xrightarrow[-N_2]{-H_2 \ (-CO_2)}$ $\xrightarrow{(-H_2)}$ (graphite)

2. Sed. Humic Acid $\xrightarrow[-N_2]{-H_2O}$ "Coal" $\xrightarrow{-H_2}$ (graphite)

3. Sed. Humic Acid $\xrightarrow[-N_2]{-H_2O}$ $\xrightarrow{(-CO_2)}$ "Kerogen (or petroleum)"

$\xrightarrow[-N_2]{-H_2}$ $\xrightarrow{-H_2}$ (graphite)

As a result of these reactions, sedimentary humic acid might convert into the substances similar to petroleum asphaltene or kerogen in sedimentary rocks, with the formation of light and heavy hydrocarbons.

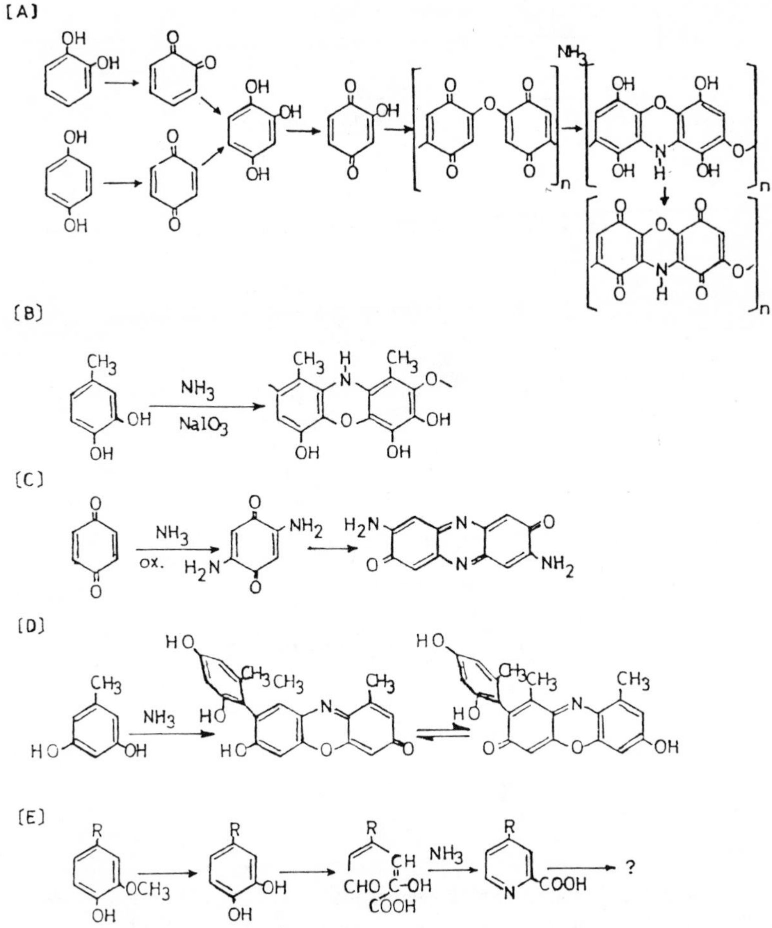

(After Bremner, Reference 50, and Flaig, Reference 51)

A

Figure 1.6   Reaction of phenolic compounds with ammonia and
their polymerization.

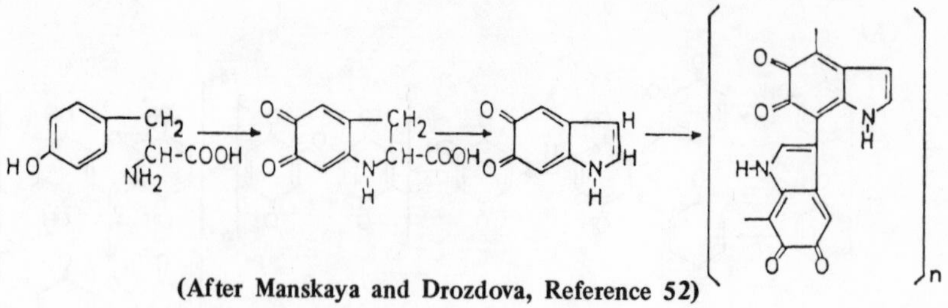

(After Manskaya and Drozdova, Reference 52)

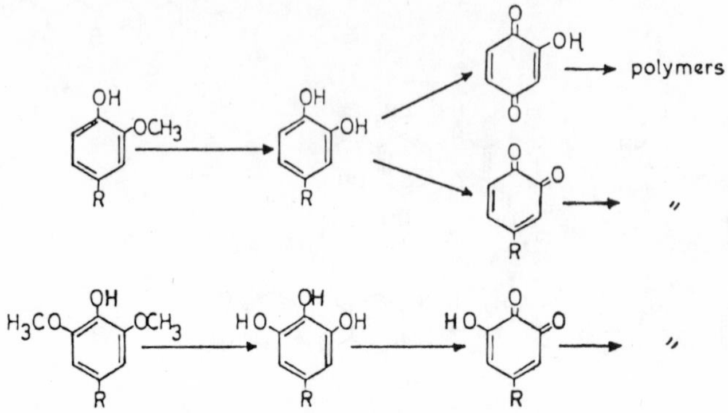

(After Flaig, Reference 53).

B

Figure 1.6. Continued

Jensen et al.[58] stated that humic acids heated to temperatures around 300°C (or higher) will, even in the presence of oxygen, tend to lose their acidic functional groups and revert to an essentially coal-like material. It is well known that coal and humic acids are very close in relationship due to their same parent material.

Yen[59] has studied the marine sediment extract by means of electron spin resonance and found that certain features are common to both humic acid and browning products when their spectra are compared (Table 1.5).

In summary, humus is important in two ways: as the mother substance for coal and oil shale, and as a transporting agent and precipitant for a number of elements. It is an important source of petroleum and the techniques to convert it to a readily utilizable form for energy will be the next step.

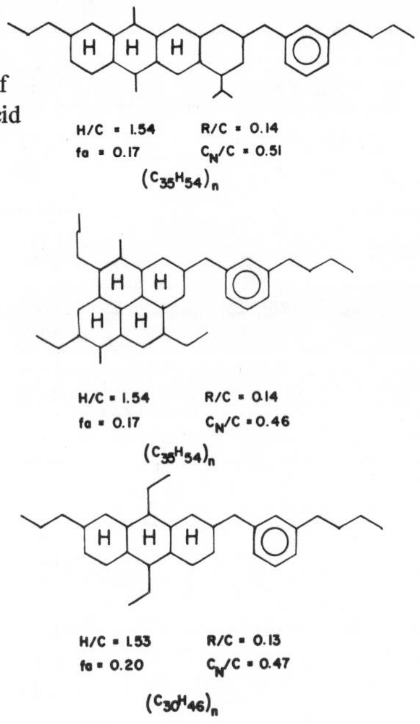

Figure 1.7 Hypothetic structural units of hydrocarbons corresponding to humic acid (H/C = 1.54).

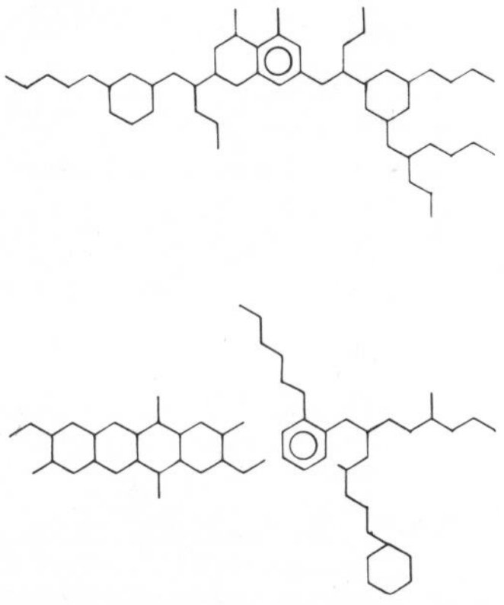

H/C = 1.54   R/C = 0.14
fa = 0.17 C_N/C = 0.51

$(C_{35}H_{54})_n$

H/C = 1.54   R/C = 0.14
fa = 0.17 C_N/C = 0.46

$(C_{35}H_{54})_n$

H/C = 1.53   R/C = 0.13
fa = 0.20 C_N/C = 0.47

$(C_{30}H_{46})_n$

Figure 1.8 Asphaltic hydrocarbons (Reference 56).

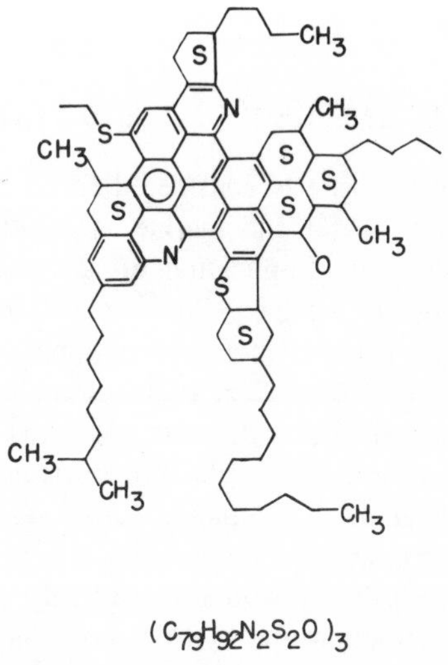

Figure 1.9 Asphaltene (Reference 57).

$(C_{79}H_{92}N_2S_2O)_3$

Table 1.5   EPR Data of Some Naturally Occurring Materials

Sample	g-Value	Width (gauss)	Shape
Chlorophyll	7.81	420	+
	5.15	500	d*
	3.26	940	+
	2.43	520	d
	2.19	60	d
Humic acid	7.99	420	−
	5.09	570	d
	2.76	1040	+
	2.15	60	d
	2.08	620	+
Browning products	7.72	440	−
	5.00	540	d
	2.90	1300	−
	2.15	60	d
	1.94	1080	+
Hemoglobin	14.95	550	d
	4.18	1750	+
	2.98	30	d
Acetone extract of	7.99	360	+
sediment	5.08	560	d
	2.73	1800	+
	2.15	20	d

*d, derivative, + and − represent upward and downward features, respectively.

MARINE KEROGEN AND PETROLEUM

Kerogen is considered to be that portion of the organic material in fresh water and marine sediments that is insoluble in ordinary solvents. Upon the application of heat, kerogen may yield gas, oil, bitumen, and organic residue. Cox[60] stated that 99% of the world's oil fields are associated with marine sediments. Known producing formations are limited to certain types of marine sedimentary rocks. Many geologists believe that sediments deposited in enclosed, or partly closed, shallow, marine basins yield more petroleum than sediments deposited in more open parts of epicontinental seas. To relate marine kerogen and energy resources, we realize that crude oil is obtained from two related sources; the primary source is evidently plant and animal lipids that supply both petroleum-type hydrocarbons and intermediates for the aromatic compounds that are present in sediments. Yen[61] indicated that the compounds formed in sediments from plant and animal intermediates such as hydrocarbons are apparently a secondary source of petroleum.

McIver[62] determined the composition of marine kerogen and concluded that kerogen is derived from varying proportions of the important naturally occurring building blocks of plants and animals; during maturation kerogen may give up commercial quantities of petroleum. He stated that any shale or fine-grained rock that contains organic matter of any kind is a potential source of hydrocarbon. If it contains a few tenths of a percent of organic matter, it is a potential source of commercial quantities of oil and gas. Figures 1.10 and 1.11 and Table 1.6 are from that study.

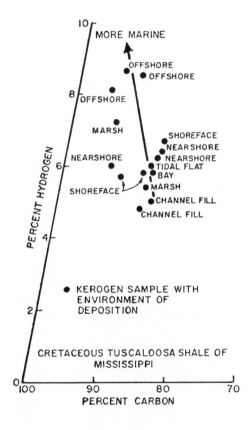

Figure 1.10   Variation of kerogen with environment of deposition (Reference 62).

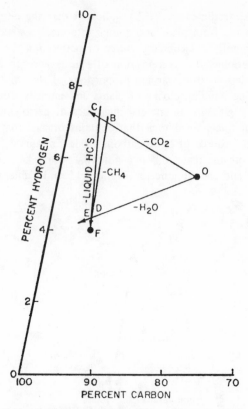

Figure 1.11 Products from maturation of kerogen (Reference 62).

Table 1.6   Products of Heating Kerogen Samples to 300°C

	CH₄ Methane (ppm)	Ethane Butane (C₂-C₄) (ppm)	Pentane Tetradecane (C₅-C₁₄) (ppm)	Total Hydro-carbon (ppm)	% CH₄	% Gases (C₁-C₄)	% Liquids (C₅-C₁₄)
Viking shale in an oil area	0.81	2.81	17.45	21.07	3.8	17.4	82.7
Atoka shale in a gas area	2.37	3.91	7.06	13.34	17.8	47.1	52.9

Sediments and crude oils contain the same types of hydrocarbons. This analogy indicates that petroleum is derived from sedimentary organic matter (Table 1.7).[63]

Table 1.7   Structural Types Occurring in Sediment and Petroleum*

I. Saturated Hydrocarbons

A. n-paraffins $CH_3(CH_2)_3CH_3$

B. Branched " $\begin{array}{c}CH_3\\CH_3\end{array}$>$CHCH_2CH_3$

C. Cycloalkanes

   1. monocyclics

   $\begin{array}{c}/CH_2\text{-}CH_2\backslash\\CH_3\text{-}CH_2 \quad\quad CH_2\\\backslash CH_2\text{-}CH_2/\end{array}$ or ⬡-CH3

   2. bicyclics

   $\begin{array}{c}CH_2\text{-}CH\text{-}CH_2\\CH_2\quad\quad CH_2\\CH_2\text{-}CH\text{-}CH_2\end{array}$

   3. tricyclics

   $\begin{array}{c}CH_3\\\backslash\\CH_2\text{-}C\text{-}CH_2\\CH\text{-}CH_2\text{-}CH\\CH_2\text{-}CH\text{-}CH_2\end{array}$

   4. tetracyclics

   R = alkyl

   5. pentacyclics

   6. hexa-to decacyclics
      structures unknown

II. Aromatics (R = CH3 etc.)

   1. alkylbenzenes

   2. cycloalkylbenzene

   3. alkylnaphthalenes

   4. cycloalkylnaphthalenes

Table 1.7, Continued

5. alkylphenanthrenes

6. cycloalkylnaphthalenes

7. pyrene

8. fluoranthene

(following compounds have been identified only in sediments)

9. triphenylene

10. 1,2 benzanthracene

11. chrysene

12. perylene

13. 1,12 benzperylene

14. coronene

*Based on Reference 63.

Studies by Hunt *et al.*[64,65] have shown that a large majority of non-reservoir sediments contain from a few hundredths to five percent of kerogen. The kerogen isolated from the marine sediments in their study was obtained as very fine, amorphous, soft powders that vary from dark brown to jet black, many showing a marked resemblance to coal dust. Refractive indices of the powdered kerogen were determined and compared with coals and asphalts as shown in Figure 1.12.

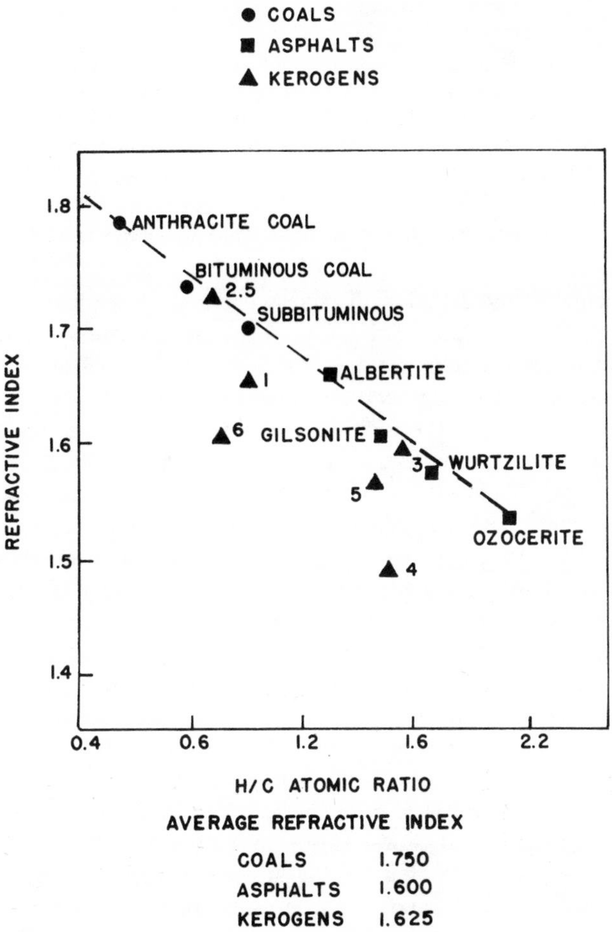

Figure 1.12 Refractive indices of coals, asphalts and kerogens.

The conclusive statements from these studies are:

1. Kerogen and petroleum are related in some way.
2. Kerogen and oil are related because of the similarity of C^{13}/C^{12} ratios.
3. Experimental results support the idea that petroleum is derived from kerogen.
4. The fact that insoluble organic matter and hydrocarbons are present in recent marine sediments as well as ancient rocks is evidence that petroleum did not originate from kerogen alone and shows that kerogen could not have been derived from petroleum.

5. Part of the insoluble organic matter in sediments is converted to hydrocarbons during diagenesis.

6. There are only two broad types of kerogen, a coal type or an oil shale type, associated with rocks that could have generated petroleum. It is hard to visualize how many different kinds of crude oils could have originated solely from these two types of kerogen.

7. Petroleum is a mixture of hydrocarbon assemblages formed by two or more independent processes in the same source bed.

8. Part of petroleum is derived from kerogen and part is of different origin.

Laphante[66] stated that with increased burial, disseminated sedimentary organic matter undergoes carbonization by processes very similar to the thermochemical reactions causing coalification. Carbonization is a thermal process marked by the generation of volatiles relatively rich in oxygen and hydrogen and by the formation of a kerogen residue, increasingly rich in carbon. The most significant oxygen-rich volatile is carbon dioxide, and the most significant hydrogen-rich volatile is hydrocarbons. With regard to the types of hydrocarbons generated during carbonization, experience has shown that gas and gas-condensate accumulations commonly are associated with low-hydrogen kerogen. The kerogen is similar to coal in composition ($< 6\%$ hydrogen) and usually found in sediments receiving mostly terrigenous organic detritus. Experience also has shown that relatively large amounts of oil-like bitumens are found in sediments containing high-hydrogen kerogen, with $> 7\%$ hydrogen[67] and commonly associated with well-defined marine and lacustrine sediments. Hypothetical chemical structures for low-hydrogen kerogen from the Gulf Coast Miocene sample and high-hydrogen kerogen from the Delaware basin Permian sample are made by Laphante and shown in Figure 1.13

The low-hydrogen content of the Miocene kerogen severely limits the high molecular weight paraffinic groups necessary for oil formation. This type of kerogen is limited mostly to gas generation by the elimination of methyl and other small hydrocarbon groups. The Permian kerogen contains sufficient hydrogen to develop liquid-hydrocarbon as well as gaseous-hydrocarbon groups. This type of kerogen has the potential for both oil and gas generation. The higher hydrogen content of the Permian kerogen also indicates a much greater total convertibility to hydrocarbons. These interpretations support similar conclusions by earlier research geochemists.

Thus, the possibilities and applications of marine kerogen to energy resources are fairly workable. According to Hunt[65] distribution of hydrocarbons and associated organic matter in recent and ancient nonreservoir sediments are summarized in Table 1.7.

Assuming that the volume of recent sediments in Gulf of Mexico is 6.3×10^{14} liters, there are 4×10^{15} g or 2×10^{12} lb of kerogen in the

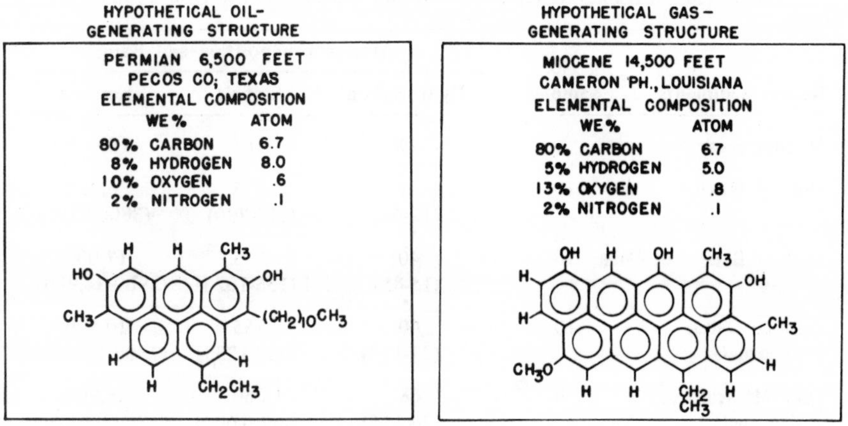

Figure 1.13   Hypothetical structures of oil- and gas-generating kerogen.

sediments (6400 ppm kerogen in Table 1.8). Using the formula $C_{215}H_{330}O_{12}N_5S$ for kerogen molecule (*Synthetic Fuels Data Handbook*, 1975), the calculation of the gross heating value may be made in accordance with Dulong's formula:

$$14,600 (C) + 61,000 (H - O/8) + 4,000 (S)$$

where 14,600, 61,000 and 4,000 represent the approximate amount of heat produced by combustion of one pound of C, H_2 and S, respectively. C, H, O and S are expressed as weight percentage. From the above, the gross heating value of pure kerogen is about 17,600 Btu/lb. Thus the amount of sediments in Gulf of Mexico will generate 3.5×10^{16} Btu of heat or 35 Quad. This is equivalent to the amount of heat generated by 20 mile^3 of coal or 7.4×10^{11} tons of fuel oil. It is a tremendous quantity of energy if we can utilize it in the future.[68]

The operating capabilities for marine kerogen will consider the exploration, drilling and production capabilities. Geophysical exploration is probably the only aspect of marine sediments development that is faster and cheaper offshore than on land. The drilling is much simpler than for oil and gas. But it still requires vast amounts of investment and baseline information, including the physical limitations such as size, shape, depth, physical characteristics and distance from shore. Production capabilities must include the development of techniques to isolate kerogen from sediments and methods to convert kerogen into oil, gas or heat for energy sources.

Table 1.8   Organic Matter in Recent and Ancient Nonreservoir Sediments

Recent Sediments	No. Samples	Average Composition and Ranges		
		Hydrocarbon	Asphalt	Kerogen
Mediterranean Sea	1	29	461	(9,000)
Gulf of Mexico	10	21 (12-63)	275 (113-790)	6,400 (3800-9100)
Gulf of Batabano, Cuba	10	40 (15-85)	575 (136-1023)	17,000 (2900-34,900)
Orinoco Delta, Venezuela	10	60 (27-110)	555 (283-1355)	10,500 (7500-14,700)
Lake Maracaibo, Venezuela	8	68 (24-116)	1,060 (266-1600)	26,900 (13,700-41,500)
Carico Trench, Venezuela	16	105 (56-352)	1,250 (224-2600)	25,000 (8100-31,200
Ancient Sediments				
Shales	791	300	600	20,100
Carbonates	281	340	400	2,160

METHANE GAS

During the decomposition of organic matter in the sediments, dissolved oxygen is removed from the interstitial water by aerobic bacteria, leading to anaerobic conditions for further decomposition. Methane is well known as a product of anaerobic decomposition of organic matter. According to Stadtman and Barker[69] and others, most biologically produced CH_4 results from anaerobic reduction of carbon dioxide or one-carbon atom compounds by hydrogen or an organic hydrogen donor. It is oxidized only under aerobic conditions and by bacteria. Thus these facts account for the increase of methane with depth and for its particularly high concentration in the very reducing sediments of the Santa Barbara Basin[70] (Table 1.9). According to early reports, there is no conclusive proof for microbial formation of gaseous hydrocarbons other than methane as a catabolic product. Davis and Squires[71] used more sensitive mass spectrometer methods to report the presence of ethane, ethylene, propane, propylene, and acetylene in addition to methane produced in sediments.

Table 1.9   Gases in Sediments (mg/l)

	Ammonia	Methane	Ethane
Santa Monica Basin			
2-26"	17.9	0.03	0.03
26-50"	49.3	0.09	0.03
50-74"	82.9	0.41	0.07
Santa Barbara Basin			
4-28"	34.2	1.18	0.0003
28-52"	68.5	18.7	0.0004
52-76"	92.0	75.0	0.0023
76-100"	117.7	140.2	0.0075
100-124"	147.8	166.5	0.0025

In the report of Rashid et al.,[72] two sediment cores were collected in the Gulf of St. Lawrence. One of the cores was collected from a depression and contained methane in the range of 5,970-14,230 ppm, in comparison to 42-106 ppm in the core of open environment. The sediments of the methane-rich core are finer and higher in organic carbon, extractable organic matter, and plant pigments. The sediments of this core also contained more diverse foraminiferal assemblages.

The results of geological investigations tend to suggest a relatively fast rate of sedimentation in the depression. As a result, in the preservation of organic compounds, anaerobic subsurface conditions develop, giving rise to high concentrations of methane through fermentative processes.

A large quantity of methane was found in recent sediments on the continental shelf of Labrador.[73] Geochemical and micropaleontological analysis of the core indicated that the gas found is pure methane, along the order of 16,000 ppm at 12 m below the sediment-water interface. Organic carbon from the Cartwright Basin is low in comparison to that of the Santa Barbara Basin.

Claypool et al.[74] stated that several hundred analyses have been made both on board the Glomar Challenger and in the laboratory on gas samples returned from the Deep Sea Drilling Project. Methane was the dominant gas in all samples, commonly amounting to more than 99% of the total. Small quantities of ethane or propane were observed in areas of high heat flow, or over a possible petroleum reservoir.

Significant quantities (40 x 10^9 cu ft/cu km) of methane can be generated in the interstitial water of deep ocean sediment where reducing conditions are initiated by rapid burial of organic matter. Comparison of carbon isotope ratios (C^{13}/C^{12}) of coexisting methane and dissolved carbonate indicates that the methane originates by bacterial CO_2 reduction. At some depth in the sediment column, depending upon the temperature and concentration, methane can exceed solubility in the interstitial water, migrate upward as a gas, and reach saturation at shallower depths. If the height of the overlying water column is greater than about 1.5 km, the gaseous methane may be converted to the solid clathrate hydrate within the uppermost (about 500 m thick) layer of sediment, where temperatures are below 20-25°C.

Stabilization of methane as a solid gas hydrate could be an important factor in the accumulation of natural gas deposits by preventing loss of gaseous methane from the sediments, allowing upward migration of gaseous methane at a pace controlled by the sedimentation rate and by producing an enrichment of gaseous methane in the zone just below the lower limit of stability of the gas hydrate.

There are three possible ways of obtaining methane from marine sediments. First, the sediment can be drilled and transported through pipeline to land. Second, the solid gas hydrate can be obtained and converted back to gaseous form or used in the solid form directly. Third, the solid and gaseous forms of methane can be transformed into liquid form for use as energy resources. Since offshore gas requires additional investment in the form of pipelines to the mainland, it is somewhat less likely to be tapped for local use. In his book *Liquefied Natural Gas*, Lom[75] gave a detailed work of LNG. Table 1.10 is the minimum work for methane liquefaction. Liquefied gas has tremendous advantages, such as being used as fuel for automotive use, for cold utilization and other cryogenic processes.

Table 1.10   Methane Liquefaction

Initial Gas Pressure (atm)	Initial Gas Temperature (K)	Temperature of Sink (K)	Minimum Liquefaction Energy	
			(kcal/kg)	(kWh/kg mol)
1	300	300	261	4.86
1	314	314	284	5.28
15	300	300	162	3.56
30	300	300	137	3.02
50	300	300	118	2.60
35	289	289	119	2.62

MINERALS

As stated by Rona,[76] plate tectonics are related to mineral resources. The concepts of continental drift and sea floor spreading provide clues to the location of economically important minerals such as oil and metals. These clues have already led to promising deposits. For example, the separation of a single ancestral continent, Pangaea, from the continents of South America and Africa resulted in a sea floor spreading from the mid-Atlantic Ridge that widens the Atlantic into an ocean. As a consequence of the deformation, the Andes Mountain chain developed by reversing the inclination of the trends. Metallic minerals that melted from the Pacific plate as it plunged under South America ascend through the overlying coastal layers and were deposited there to form the metal-bearing provinces of the Andes. Meanwhile, salt, which originated in thick layers of rock salt that have been buried under sediments in continental margins, rises in large dome-shaped masses that act to trap oil and gas generated from the organic matter preserved in the former Atlantic Sea (Figure 1.14).

An adequate, dependable, and continuing supply of minerals is required to maintain industrial strength and energy resources. Fissing of one pound of uranium produces heat energy equivalent to 5900 barrels of crude oil. Copper, coal, manganese, platinum, thorium and yttrium are all energy resources.

The world's oceans are the storehouse of a wide array of minerals with greatly varying characteristics and occurrences, including those accumulated on the ocean floor and those locked in the rocks beneath the ocean floor. The following information is from "Marine Resources and Legal Political Arrangement for their Development":[77] (Table 1.11).

Barite: Uses—weighting agent in oil and gas well drilling fluids; a high density aggregate in concrete for nuclear shielding. Potential from marine sediments—a few barite nodules have been found on the continental shelves; it occurs in large quantities in marine sediments of the eastern Pacific Ocean.

Bromine: Uses—the main use of bromine is in additives for gasoline. Potential—resource potential is virtually unlimited.

Coal: Uses—most bituminous coal is burned for heat and power production. Anthracite coal is used as an energy fuel and as a source of industrial carbon. Lignite is used as a fuel and for generating electrical power. Potential—coal deposits possibly in New England, Oregon and Washington. In Alaska, coal-bearing formations project into the continental shelves in several areas, and the potential for the presence of offshore coal resources is promising.

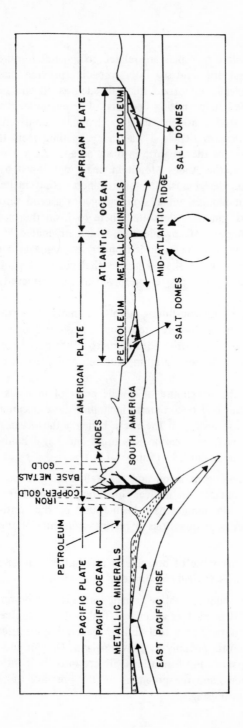

Figure 1.14 The role of plate boundaries in the accumulation of mineral deposits.

Table 1.11   Summary Table Showing Potential of Hard Mineral Commodities from U.S. Marine Sources (Including Water Column and Continental Shelves and Slopes)

Currently Produced	Good Potential	Fair to Poor Potential
Bromine	Gold	Barite
Magnesium & magnesium salts	Phosphate	Chromite
	Monazite (thorium) and rare-earths	Chromium
Oyster shell and calcium carbonate mud		Coal
	Staurolite	Lead
Salt	Kyanite and sillimanite	Mercury
Sand and gravel	Titanium minerals	Platinum
Sulfur	Zircon	Silver
		Tin
		Manganese
		Nickel, Cobalt
		Copper, Molybdenum
		Uranium
		Vanadium

Copper: Uses—about 50% of all copper consumed is for electrical applications. Potential—copper is known to occur in significant amounts in manganese nodules and on the deep sea floor of the Pacific.

Platinum Group Metals: Uses—metals of the platinum group include platinum, iridium, osmium, rhodium, and ruthenium. Major uses include catalysts in the production of high-octane gasoline. Potential—platinum has been produced in the United States from one placer area at Goodnews Bay, Alaska and as a by-product in the mining and refining of other metals, notably gold.

Thorium: Uses—thorium will become an important nuclear fuel as it can be converted in a nuclear reactor to U^{233} which is fissionable and can sustain a chain reaction. Potential—monazite, a rare earth phosphatic mineral containing up to 12% ThO_2, is the major source of the world's thorium. An estimate of total monazite production between 1895 and 1962 from southeastern states' beaches and alluvial deposits is 4,000 to 10,000 tons. The ThO_2 content of these monazites averaged only 3.5 to 4%.

Yttrium: Uses—the potential commercial uses as in nuclear reactors. Potential—yttrium is always found in association with rare earth. It is concentrated mainly in xenotime (essentially an yttrium phosphate) and

to a lesser extent in monazite and other minerals. The potential for
yttrium production depends on the abundance of xenotime and monazite
in black sands.

Zircon: Uses—physical properties of zirconium have led to its extensive use in nuclear reactors. The power plants of a number of naval
vessels contain zirconium. Potential—zircon, the sole source of zirconium,
is highly resistant and commonly concentrated in beach, dune, and
alluvial sands with rutile, ilmenite, magnetite and monazite.

Baturin *et al.*[78] summarized their work as follows:

(a) In many cores from the Black and Mediterranean Seas, there is a
similarity in the distribution of metals (U, Mo, Co, Ni, V) and of
organic matter.

(b) In the Black Sea cores the correlation between organic matter and
metals is particularly characteristic of uranium and molybdenum.

(c) The character of correlation between the metals and C_{org} in the
cores is different from that in the surface layer of the sediments.
This indicates redistribution of the metals within the **sediments**
during diagenesis.

(d) Transformation of organic matter is the principal factor influencing
the mobility of metals in the sediments.

(e) The diagenetic redistribution of molybdenum, cobalt, nickel and
possibly vanadium is related in part to the formation of pyrite
and iron-manganese concretions.

(f) Uranium is accumulated in the sediments rich in organic matter
during sedimentation and during diagenesis.

(g) Migration of uranium in the sediments may be related to the
migration of the bituminous part of organic matter.

(h) **Concentration** of uranium in organic matter of the Black Sea
sediments is related to the rate of sedimentation.

There are different degrees of potential for each mineral considered as
a possible energy resource in marine sediments. Mining problems are related mostly to motion on surface platforms, excavation, control, materials handling, and benefication. Marine corrosion and biological fouling
are important aspects that will become more obvious problems as operations are developed. Basic ideal assumptions for marine sediments would
be: that they are located near the offshore (*e.g.,* within 1 mile); that
the combined depth of water and sediments not be deep (*e.g.,* within
100 feet); and that the deposit be amenable to standard bucket line
dredging methods, giving a high recovery factor (*e.g.,* about 80%).

Only a very small amount of exploration for minerals has been done
on the continental shelves. Too little is known about the overall geology
and mineral potential of the shelves and slopes to induce industry to
spend significant sums of money on exploration in the marine environment.

However, given specific target areas that might be pointed out by geological and geophysical surveys, by more efficient exploration equipment, and by a favorable legal environment in which to operate, the larger mining companies might undertake offshore exploration on a significant scale.

New technology, as a variety of sensors that could be used to make *in situ* engineering measurements relevant to the strength and behavior of marine sediments, geochemical parameters, and physical properties of sediments at depth, would certainly help to obtain the resources from marine sediments (Table 1.12).

Finally, the economic value of the heavy metal, bound to the suspended matter and dissolved in the water, that is lost each year to the sea is worth mentioning. For example, in the Rhine alone, a total of 35,000 tons of zinc, 4000 tons of copper and lead, 500 tons of cadmium and 150 tons of mercury are discharged into the North Sea.[79] The tonnage of zinc, mercury and cadmium corresponds to 0.8%, 1.5% and 3%, respectively, of the annual world production of these metals. Of the world's lead production, 15% is consumed for gasoline additives; during combustion almost the entire lead content is emitted into the ambient air. At least 20% of the world's mercury production is employed in chlor-alkali-electrolysis; a large amount is released into surface waters.[79]

CONCLUSION

One of the most important theoretical problems of petroleum geology is that of the origin of petroleum. Studies of the organic matter of recent marine sediments and of the initial stages of its formation are important elements in the solution of this problem.

The discovery of liquid hydrocarbons in recent marine sediments has demonstrated that main components of petroleum are already present in sediments at the earliest stages of diagenesis. By studying the organic matter of various types of sediments and the way its composition is affected in transforming into the fossil state, it should be possible to determine the distinguishing features of petrolenous rocks and to establish which conditions are most favorable to the formation of petroleum.

In the above discussion we have purposely omitted deuterium and tritium, which are evidently very important from a fusion energy point-of-view. As reserves for the concentrated form are depleting, technology and science accordingly will shift to explore the more dilute form.

Table 1.12 U.S. and World Land Reserves and Resources of Some Metals and Nonmetals that May Occur in the Continental Shelves[a]

Commodity	Minable Reserves[b]		Resources[c]		Projected Cumulative Demand, 1966-2000 (Rounded from Bureau of Mines Estimates)	
	U.S.	World (including U.S.)	U.S.	World (including U.S.)	U.S.	World (including U.S.)
Aluminum (bauxite, millions, long tons)	45	5,800	300	9,600	440[e]	840
Barite (millions, short tons)	60	130	100	d	80	190
Beryllium (short tons, equiv. beryl)	d	d	1,000,000	1,650,000	360,000	540,000
Bromine (million pounds)	Vast	Vast	Vast	Vast	22,000	30,000
Borates (millions, short tons B$_2$O$_3$)	95	110	d	Several billion	4	14
Chromium (millions, long tons, chromite)	0	2,000	8	d	22	74
Copper (millions, short tons)	86	210	65	d	140[e]	400
Cobalt (thousands, short tons)	50	2,200	3	d	600	1,300
Gold (millions, troy ounces)	50	1,000	400	d	670	2,370
Helium (billions, cubic feet)	154	d	42	d	60	d
Industrial diamonds (million carats)	0	d	0	d	1,700	3,600
Iron ore (millions, long tons)	8,000	250,000	100,000	250,000	6,400	35,000
Lead (millions, short tons)	35	83	15	d	57	180
Manganese ore (millions, long tons)	0	3,800	1,000	15,000	50	450
Mercury (thousands of flasks)	200	7,000	500	10,000	3,600[e]	11,400
Nickel (thousands, short tons)	250	60,000	1,400	d	14,500[e]	31,700
Niobium (thousands, short tons, Nb$_2$O$_5$)	125	9,800	165	8,600	190	370
Phosphate (millions, long tons)	12,000	48,000	48,000	d	500	2,000

Potash (millions, short tons, K_2O)	1,400	72,000	5,000	Very large	300	1,100
Platinum Group (millions, troy ounces)	d	280	d	d	144[e]	250
Rare Earths (millions, short tons, Re_2O_3)	5	d	d	5	0.5	1
Salt (trillions, short tons)	60	Vast	d	d	0.003	0.009
Silver (millions, troy ounces)	1,400	5,500	500	d	14,000[e]	31,000
Sulfur (millions, short tons)	d	d	500	2,000	690	2,600
Tantalum (short tons, Ta_2O_5)	2,100	170,000	d	95,000	54,000	104,000
Tin (thousands, long tons)	9	5,600	43	11,400	3,200[e]	9,200
Titanium (millions, short tons, TiO_2)	100	500	d	Vast	50	100
Thorium (thousands, short tons, ThO_2)	0	82	200	1,000	5	15
Tungsten (thousands, short tons)	70	1,500	200	d	620	2,040
Uranium (thousands, short tons, U_3O_8)	210	742	675	2,700	1,500	3,750
Vanadium (thousands, short tons)	200	3,500	1,300	20,000	650	1,000
Zinc (millions, short tons)	29	100	60	d	90	280
Zircon (millions, short tons)	6	30	d	d	3.0	10

aFrom Reference 73.

bMinable reserves are materials that may or may not be completely explored but that may be quantitatively estimated and are considered to be economically exploitable at the time of the estimate.

cResources are materials other than reserves that are prospectively usable and include undiscovered recoverable resources as well as those whose exploration requires more favorable economic or technologic conditions.

dUnknown

eDemand figures include a significant quantity of recycled metal.

ACKNOWLEDGMENTS

Partial support from the following sources is acknowledged: AEC E(29-2)-3619; A.G.A. BR-48-12; NOAA-Sea Grant R/RD-2; and ACS-PRF 6272 AC2. The authors would like to express their thanks also to Won Wook Choi for the technical input to a portion of this paper.

REFERENCES

1. Goldberg, E. D. (Ed.) *The Sea, Vol. 5: Marine Chemistry* (New York: John Wiley & Sons, 1974).

2. Yasushi, Kitano (Ed.) *Geochemistry of Water*, Benchmark Papers in Geology, Vol. 16, Dowden, Hutchingson & Ross, Inc. (1975).

3. Riley, J. P. and G. Skirrow (Ed.) *Chemical Oceanography*, Vol. 3, 2nd ed. (New York: Academic Press, Inc., 1975).

4. Gross, M. G. *Oceanography—A View of the Earth* (Englewood Cliffs, New Jersey: Prentice-Hall, Inc., 1972).

5. Pytkowicz, R. M. "Some Trends in Marine Chemistry and Geo-chemistry," *Earth Science Rev.* 11, 1-46 (1975).

6. Bathurst, R. G. C. *Carbonate Sediments and Their Diagenesis*, Developments in Sedimentology Series No. 12, 2nd enlarged ed. (Amsterdam: Elsvier Scientific Publishers, Inc., 1975).

7. Stumm, W. and J. J. Morgan. *Aquatic Chemistry* (New York: Wiley—Interscience, 1970).

8. Siever, R. "Sedimentological Consequences of a Steady-State Ocean Atmosphere," *Sedimentology*, 11, 5-29 (1968).

9. Garrels, R. M. and F. T. Mackenzie. *Evaluation of Sedimentary Rocks* (New York: Norton, 1971).

10. Yen, T. F. "Terrestrial and Extraterrestrial Stable Organic Mole-cules," in *Chemistry in Space Research* (New York: Elsvier Scientific Publishers, 1972), pp. 105-153.

11. Rashid, M. A. "Contribution of Humic Substance to the Cation Exchange Capacity of Different Marine Sediments," *Marine Sed.* 5, 44 (1969).

12. Chen, K. Y. and T. F. Yen. "Models for the Fate of Heavy Metals in Sediments," *ACS Div. Water, Air and Waste Chemistry Priprints* 12, 165-179 (1972).

13. Scott, R. B., P. A. Rona, L. W. Butler, A. S. Nawalk and M. R. Scott. "Manganese Crust in Atlantic Fracture Zone," *Nature* (London) 239, 77-79 (1972).

14. Glassby, G. P. "The Mineralogy of Manganese Nodules from a Range of Marine Environments," *Marine Geol.* 13, 57-72 (1972).

15. Heimendahl, M. V., G. L. Hubred, D. W. Fuerstenau and G. Thomas. "A Transmission Electron Microscope Study of Deep-Sea Manganese Nodules," *Deep-Sea Res.* 23, 69-79 (1976).

16. Volkovic, V. *Trace Element Analysis* (London: Taylor and Francis Ltd., 1975).

17. Church, T. M. and M. Bernat. "Thorium and Uranium in Marine Barite," *Earth Planet. Sci. Lett.* **14**, 139-144 (1972).

18. Bernat, M., R. H. Bjeri, M. Koide, J. Griffin and E. D. Goldberg. "Uranium, Potassium and Argon in Marine Phillipsites," *Geochim. Cosmochim. Acta* **34**, 1053-1071 (1970).

19. Veeh, H. M., S. E. Calvert, and N. B. Price. "Accumulation of Uranium in Sediments and Phosphorites on the South West African Shelf," *Marine Chem.* **2**, 189-202 (1974).

20. Yen, T. F. "Heavy Metals in Sediments," Gordon Research Conference on Organic Geochemistry, Hampton (1972).

21. Perry, E. A., Jr., J. M. Gieskes and J. R. Lawrence. "Mg, Ca and O^{18}/O^{16} Exchange in the Sediment-Pore Water System, Hole 149, DSDP," *Geochim. Cosmochim. Acta* **40**, 413-423 (1976).

22. Bischoff, J. L. and T. L. Ku. "Pore Fluids of Recent Marine Sediments. II. Anoxic Sediments of $35°$ to $45°$ N Gibraltar to Mid-Atlantic Rdige," *J. Sediment. Petrol.* **41**, 1008-1017 (1971).

23. Sayles, F. K. and F. F. Manheim. "Interstitial Solutions and Diagenesis in Deeply Buried Marine Sediments: Results from the Deep Sea Drilling Project," *Geochim. Cosmochim. Acta* **39**, 103-128 (1975).

24. Duchart, P., S. E. Calvert and N. B. Price. "Distribution of Trace Metals in the Pore Waters of Shallow Water Marine Sediments," *Limnol. Oceanog.* **18**(4), 605-610 (1973).

25. Chen, K. Y., S. K. Gupta, A. Z. Sycip, J. C. S. Lu, M. Knezevic and W. W. Choi. "Research Study on the Effect of Dispersion, Settling and Resedimentation on Migration of Chemical Constituents during Open Water Disposal of Dredged Materials," U.S. Army Engineer Waterways Experimental Station, Vicksburg, Miss., Contract No. DACW39-74-C-007 (December 1975).

26. Lu, J. C. S. "Studies on the Long-Term Migration and Transformation of Trace Metals in the Polluted Marine Sediment-Seawater System," Ph.D. Thesis, University of Southern California (June 1976).

27. Sayles, F. L., T. S. R. Wilson, D. N. Hume and P. C. Mangelsdorf, Jr. "*In situ* Sampler for Marine Sedimentary Pore Waters: Evidence for Potassium Depletion and Calcium Enrichment," *Science* **181**, 154-156 (1973).

28. Duursma, E. K. and C. J. Bosch. "Theoretical, Experimental and Field Studies Concerning Diffusion of Radioisotopes in Sediments and Suspended Particles of the Sea, Part B: Method of Experiments," *Neth. J. Sea Res.* **4**, 395-469 (1970).

29. Lerman, A. "Maintenance of Steady-State in Oceanic Sediments," *Amer. J. Sci.* **275**, 609-635 (1975).

30. Lerman, A. "Uptake and Migration of Tracers in Lake Sediments," *Limnol. Oceanog.* **20**(4), 497-510 (July 1975).

31. Li, Y. H. and S. Gregory. "Diffusion of Ions in Sea Water and in Deep-Sea Sediments," *Geochim. Cosmochim. Acta* **38**, 703-714 (1974).

32. Hurd, D. C. "Factors Affecting Solution Rate of Biogenic Opan in Seawater," *Earth Planet. Sci. Lett.* **15**, 411-417 (1972).

33. Feick, G., R. A. Hoyne and D. Yeaple. "Release of Mercury from Contaminated Freshwater Sediments by the Runoff of Road De-Icing Salt," *Science* **175**, 1142-1143 (1972).
34. Rashid, M. A. "Role of Humic Acids of Marine Origin and their Different Molecular Weight Fractions in Complexing Di- and Tri-Valent Metals," *Soil Sci.* **111**, 298-306 (1971).
35. Leonard, R. G. "Organic versus Inorganic Trace Metal Complexes in Sulfide Marine Waters—Some Speculative Calculations based on Available Stability Constants," *Geochim. Cosmochim. Acta* **33**, 1297-1302 (1974).
36. Faust, S. J. and J. V. Hunter. *Organic Compounds in Aquatic Environments* (New York: Marcel Dekker, Inc., 1971).
37. Gjessing, E. T. *Physical and Chemical Characteristics of Aquatic Humus* (Ann Arbor, Michigan: Ann Arbor Science Publishers, 1976).
38. Serruya, C., M. Edelstein, U. Pollingher and S. Serruya. "Lake Kinneret Sediments: Nutrient Composition of the Pore Water and Mud-Water Exchanges," *Limnol. Oceanog.* **17**(1) (January 1972).
39. Harrison, M. J., R. E. Pacha, and R. Y. Morita. "Solubilization of Inorganic Phosphates by Bacteria Isolated from Upper Klamath Lake Sediment," *Limnol. Oceanog.* **17**(1) (January 1972).
40. Li, W. C., D. E. Armstrong, J. D. H. Williams, R. F. Harris and J. K. Syers. "Rate and Extent of Inorganic Phosphate Exchange in Lake Sediments," *Proceedings Soil Sci. Soc. Amer.* **36**, 279-285 (1972).
41. Barnes, R. O., K. K. Bertin and E. D. Goldberg. "N_2: Air, Nitrification and Denitrification in Southern California Borderland Basin Sediments," *Limnol. Oceanog.* **20**(6), 962-970 (November 1975).
42. Austin, E. R. and G. G. Lee. "Nitrogen Release from Lake Sediments," *J. Water Poll. Control Fed.* **45**(5), 870-879 (1973).
43. Graetz, D. A., D. R. Keeney and R. B. Aspiras. "Eh Status of Lake Sediment-Water Systems in Relation to Nitrogen Transformations," *Limnol. Oceanog.* **18**(6), 285-295 (1976).
44. Vanderborght, J. P. and G. Billen. "Vertical Distribution of Nitrate Concentration in Interstitial Water of Marine Sediments with Nitrification and Denitrification," *Limnol. Oceanog.* **20**(6), 953-961 (November 1975).
45. Skopintsev, B. A. "Organic Matter in Natural Water (Water Humus)," *Tr. Geol. Inst.*, Akad. Nauk S.S.S.R. **17**(29) (1950).
46. Khan, D. V. "Concerning the Connection Between Organic Matter and Soil Minerals," *Dokl. Akad. Nauk. S. S. S. R.* **31**(3) (1951).
47. Bordovskiy, O. K. "Accumulation and Transformation of Organic Substances in Marine Sediments," *Marine Geol.* **3**, 1-83 (1964).
48. Young, D. K., S. R. Sprang and T. F. Yen. "Preliminary Investigation on the Precursors of the Organic Components in Sediments—Melanoidin Formations," Chapter 5 this volume.
49. Degens, E. T., J. H. Euter and N. F. Shaw. "Biochemical Compounds in Offshore California Sediments and Sea Water," *Geochim. Cosmochim. Acta* **28**, 45-66 (1963).

50. Bremer, J. M. "Nitrogenous Compounds," in *Soil Biochemistry*, A. D. McLaren and G. H. Peterson, Ed. (New York: Marcel Dekker, Inc., 1967), pp. 19-66.

51. Flaig, W. "Origin of Nitrogen in Coals," *Chem. Geol.* 3, 161-187 (1968).

52. Manskaya, S. M. and T. V. Drozdova. *Geochemistry of Organic Substances*, L. Shapiro and I. Breger, trans. (New York: Pergamon Press, 1968).

53. Flaig, W. "Chemistry of Humic Substances in Relation to Coalification," in *Coal Science*, R. F. Gould, Ed. (American Chemical Society, 1966).

54. Ishiwatari, R. "Organic Polymers in Recent Sediments–Chemical Nature and Fate in Geological Environment," A Thesis for the degree of Doctor of Science, Tokyo Metropolitan University, Tokyo Japan (1971).

55. Ishiwatari, R. "Chemical Nature of Sedimentary Humic Acids," *Proc. Int. Meeting Humic Substances* (Nieuwersluis, 1972), pp. 87-107.

56. Wetmore, D. E., C. K. Hancock and R. N. Traxler. "Fractionation and Characterization of Low Molecular Weight Asphaltic Hydrocarbons," *Anal. Chem.* 38, 225-230 (1966).

57. Winniford, R. S. and M. Bersohon. "The Structure of Petroleum Asphaltenes as Indicated by Proton Magnetic Resonance," Symp. Tars, Pitches, Asphalts, *Am. Chem. Soc., Div. Fuel Chem., Preprints* 1962, 21-32 (through Witherspoon, P. A. and Winniford, R. S.) (1967).

58. Jensen, E. J., N. Melnyk, J. C. Wood and N. Berkowitz. "The Dry Oxidation of Subbitumenous Coal," in *Coal Science*, R. F. Gould, Ed. (American Chemical Society, 1966).

59. Yen, T. F. Unpublished Results.

60. Cox, B. B. "Transformation of Organic Material into Petroleum under Geological Conditions," *Bull. Am. Assoc. Petrol. Geol.* 30, 645-659 (1946).

61. Yen, T. F. "Genesis and Degradation of Petroleum Hydrocarbons in Marine Environments," ACS Symposium Series, No. 18, (1975), pp. 231-266.

62. McIver, R. D. "Composition of Kerogen–Clue to Its Role in the Origin of Petroleum," *Seventh World Petroleum Congress, Proceedings*, Vol. 2 (1967), pp. 25-36.

63. Meinschein, W. G. "Origin of Petroleum," *Bull. Am. Assoc. Petrol. Geol.* 43, 925-943 (1959).

64. Hunt, J. M. and G. W. Jamieson. "Oil and Organic Matter in Sources Rocks of Petroleum," *Bull. Am. Assoc. Petrol. Geol.* 40, 477-488 (1956).

65. Hunt, J. M. "Distribution of Hydrocarbons in Sedimentary Rocks," *Geochim. Cosmochim. Acta* 22, 37-49 (1961).

66. Laphante, R. E. "Hydrocarbon Generation in Gulf Coast Tertiary Sediments," *Am. Assoc. Petrol. Geol. Bull.* 58, 1281-1289 (1974).

67. Smith, J. M. "Conversion Constants for Mahogany-Zone Oil Shale," *Am. Assoc. Petrol. Geol. Bull.* 50, 167-170 (1954).

68.   Weeks, L. G. "Petroleum Resources Potential of Continental Margins," *Geology of Continental Margins*, C. A. Burk and C. L. Drake, Eds. (New York: Springer-Verlag, 1974), pp. 953-964.

69.   Stadtman, T. C. and H. A. Barker. "Studies on the Methane Fermentation, VII and IX," *J. Bacteriol.* **61**, 67-86 (1951).

70.   Emery, K. O. and D. Hoggan. "The Gases in Marine Sediments," *Am. Assoc. Petrol. Geol. Bull.* **42**, 2174-2188 (1958).

71.   Davis, J. B. and R. M. Squires. "Detection of Microbially Produced Gaseous Hydrocarbons Other than Methane," *Science* **119**, 381-382 (1954).

72.   Rashid, M. A., G. Vilks and J. D. Leonard. "Geological Environment of a Methane-Rich Recent Sedimentary Basin in the Gulf of St. Lawrence," *Chem. Geol.* **15**, 83-96 (1975).

73.   Vilks, G., M. A. Rashid and W. J. M. Van der Linden. "Methane in Recent Sediments of the Labrador Shelf," *Can. J. Earth Sci.* **11**, 1427-1434 (1974).

74.   Claypool, G. E. "Generation of Light Hydrocarbon Gases in Deep-Sea Sediments," *Am. Assoc. Petrol. Geol. Bull.* **57**, 773

75.   Lom, W. L. *Liquified Natural Gas.* (New York: John Wiley & Sons, 1974).

76.   Rona, P. A. "Plate Tectonics and Mineral Resources," *Sci. Amer.* **229**(86) (1973).

77.   Superintendent of Documents. "Marine Resources and Legal-Political Arrangements for their Development," U.S. Gov. Printing Office, Humus Substances in Sediments.

78.   Baturin, G. N., A. V. Kochenov and K. M. Shimkus. "Uranium and Rare Metals in the Sediments of the Black and Mediterranean Seas," *Geokhimiya* **1**, 41-50 (1966).

79.   Forstner, U. and G. Muller. "Heavy Metal Accumulation in River Sediments: A Response to Environmental Pollution," *Geoforum* **14**, 53-61 (1973).

80.   Bortelson, G. C. and G. F. Lee. "Phosphorous, Iron and Manganese Distribution in Sediment Cores of Six Wisconsin Lakes," *Limnol. Oceanog.* **19**(5), 794-801 (1974).

2

CHARACTERIZATION OF SEDIMENTS IN THE VICINITY OF OFFSHORE PETROLEUM PRODUCTION

R. M. Bean, J. W. Blaylock,
E. A. Sutton and R. E. Wildung

Ecosystems Department
Battelle, Pacific Northwest Laboratories
Richland, Washington 99352

F. M. Davidson

Petroleum Analytical Research Corporation
Houston, Texas 77012

INTRODUCTION

In view of the increasing dependence on offshore drilling to satisfy the world's energy needs, it is evident that means must be developed and implemented for determining the effects of oil on marine environments. In particular, sediments in the vicinity of oil production operations need to be studied, since this environmental compartment represents a potential repository for nonvolatile, insoluble residues of discharged oil. As a first step in determining the relative contributions of residues from petroleum production, industrial and domestic effluents, and natural detritus to sediments, procedures are needed for characterization of sediment organic compound types.

Characterization of the organic fraction of sediments is often made difficult by the low availability of material for analysis; *i.e.,* on the order of a few hundred micrograms per gram of dried sediment. Low concentrations of organic material restrict the use of physical properties as a characterization tool. Gas chromatography has been used to characterize sediments in the vicinity of spills of distillate fuels,[1] but both qualitative and quantitative gas chromatographic analyses are dependent on the volatility of the organic extract, and are thus of limited utility where bituminous residues predominate. Quantitative analysis of total bitumen in sediments using infrared spectral measurements has been reported,[2] but provided little information with respect to the compound types present. Thin layer chromatography has been applied recently to the characterization of sediment extracts according to distribution of compound types,[3] but the results are best characterized as semiquantitative.

The offshore production area of study was Lake Maracaibo, in the state of Zulia, Venezuela. This body of water is 150 km long and 110 km at maximum width, with an average depth of 25.9 meters (Figure 2.1). Active petroleum production in the northeast quadrant of the lake has been under way for more than 40 years. Current daily production of oil is approximately 2.5 million barrels. In addition to being subject to oil spills occurring during the course of production operations, the lake is also contaminated by natural oil seeps. The lake is a dynamic system of brackish water that alternates between incursion of seawater through the Straits of Maracaibo from the north in the dry season, and a general flushing action during the rainy season from large river systems in the south. The geography, hydrography and general ecology of Lake Maracaibo has been reported in a recent study.[4] Because of the localization of oil production in the Northeast, an opportunity was afforded to compare the organic composition of sediments taken from the producing zone with those from areas remote from production operations.

Fractionation of extractable organic material from sediment samples was accomplished using an adaptation of methodology previously reported by Koons and Monaghan,[5] developed to determine the organic composition of drill cores. The fractions were analyzed for carbon, hydrogen, nitrogen, sulfur and molecular weight when sufficient material was available. High-voltage mass spectrometric methods were applied to saturate and aromatic fractions to determine the compound-type distributions present in these fractions. Statistical approaches were utilized in interpreting the analytical results.

METHODS

Sampling

Two sampling programs were undertaken. In an initial screening phase, the top 5 cm of core samples from 20 stations were taken to determine the extent of variability in composition throughout the region under study and to investigate trends relating the character of the organic fractions to

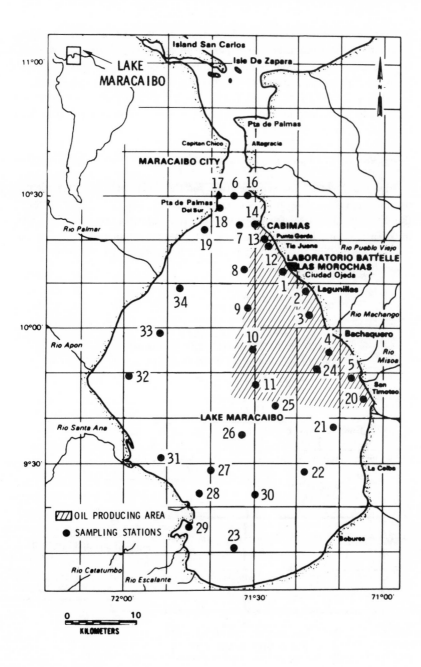

Figure 2.1   Map of Maracaibo showing oil-producing areas.

geographic location. Duplicate cores from a few locations were analyzed to estimate the variability in sampling. In addition to the core samples, larger samples were taken with an Ekman dredge (approximately 0-10 cm) from 15 stations. Analyses of the organic constituents of the dredge samples were performed in triplicate. Both types of samples were frozen at the site of collection and maintained in a frozen state until they were freeze-dried prior to analysis.

Separation of Sediment Organic Fractions

Figure 2.2 shows the scheme used for the fractionation of sediment organic material. The dried sediment (10-50 g) was placed in a Pyrex® thimble with a coarse, fritted-glass filter and Soxhlet-extracted with 500 ml of a constant boiling mixture of benzene and methanol (3:2) for 16 hours. Extraction for longer periods did not significantly increase the quantity of material extracted.

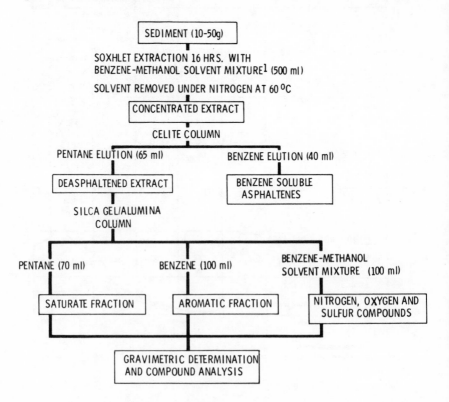

SEDIMENT (10-50g)

SOXHLET EXTRACTION 16 HRS. WITH
BENZENE-METHANOL SOLVENT MIXTURE[1] (500 ml)

SOLVENT REMOVED UNDER NITROGEN AT 60 °C

CONCENTRATED EXTRACT

CELITE COLUMN

PENTANE ELUTION (65 ml)   BENZENE ELUTION (40 ml)

DEASPHALTENED EXTRACT   BENZENE SOLUBLE ASPHALTENES

SILCA GEL/ALUMINA COLUMN

PENTANE (70 ml)   BENZENE (100 ml)   BENZENE-METHANOL SOLVENT MIXTURE (100 ml)

SATURATE FRACTION   AROMATIC FRACTION   NITROGEN, OXYGEN AND SULFUR COMPOUNDS

GRAVIMETRIC DETERMINATION AND COMPOUND ANALYSIS

Figure 2.2   Scheme for analysis of sediments.

The organic extract was evaporated to dryness under a stream of nitrogen at 60°C, redissolved in benzene and filtered through a glass-wool plug to remove salts. The filtrate was evaporated to dryness at room temperature and the dry weight determined. This material was designated as the total organic extract.

The total organic extract was treated with 25 ml of pentane for one hour to precipitate the asphaltenes. The solution, containing pentane-soluble material, was transferred to a 30 x 1-cm column packed with 15-20 cm of Fisher Hyflo Super-Cel® prewashed with pentane. The column was eluted with 40 ml of pentane and the eluate was transferred to a beaker. The eluate was concentrated by evaporation at room temperature to approximately 5 ml, transferred to a tared aluminum weighing pan, evaporated to dryness, and the dry weight determined. This material was taken as the deasphaltened extract.

The residue remaining in the beaker after removal of pentane solubles was dissolved in 10 ml benzene, transferred to the column, and eluted with an additional 30 ml benzene. The eluate was concentrated at 60°C to approximately 5 ml, transferred to a tared aluminum weighing pan, evaporated to dryness; and the soluble material, designated as benzene-soluble asphaltenes, was determined gravimetrically.

The asphaltene-free extract, dissolved in 10 ml pentane, was transferred to another 30 x 1-cm column packed with 10 g Davison Grade 923 silica gel activated at 150°C for 16 hours under 10 g Alcoa F-20 alumina activated at 400°C for 16 hours. The column was eluted with a total of 70 ml of pentane. The total eluate was concentrated, dried, and the concentration of soluble material determined gravimetrically as previously described for the pentane extract. This material was designated as the saturate fraction.

The column containing the residual asphaltene-free extract was eluted with 100 ml of benzene, and the concentration of soluble material in the eluate determined as described previously for the benzene extract. This material was taken as the aromatic fraction.

The column containing the remaining asphaltene-free extract was finally eluted with 100 ml of a 1:1 solution of methanol and benzene. The material in this fraction taken to contain nitrogen, sulfur and oxygen compounds (polar or NSO fraction) was determined as described for the benzene extract.

Mass Spectrometry

Sediment saturate and aromatic fractions were drawn into a glass capillary, the ends of the capillary sealed, and the sample placed in the inlet of a CEC-103 mass spectrometer. The capillary was broken with a striker in the inlet system, which was maintained at 305°C and constant vacuum (10^{-6} mm Hg). An ionizing voltage of 70 V was applied, and the spectrum was recorded using an Infotronics digital readout system with a high-speed printer.

RESULTS AND DISCUSSION

Recovery Efficiency and Reproducibility of Sediment Extraction Procedure

To evaluate the effectiveness of the fractionation procedure as a characterization tool, the recovery of bituminous material from sediments was determined. A sample of bitumen was prepared by weathering a thin film of petroleum, obtained from the producing area, in a water bath at 60°C for 64 hours. The prepared bitumen (50 mg) was fractionated according to the above procedure. An equivalent amount of bitumen was added to sediment (22 g), which had been previously extracted with benzene—methanol and vacuum dried. Table 2.1 shows the results of this study. Recoveries from the sediment based on the sediment-free bitumen analysis are acceptable at the level investigated, with standard deviations of less than 10% except for the polar fraction. Lower recoveries of benzene-soluble aslphaltenes may reflect a higher degree of association of these polar molecules with the sediment.

Table 2.1   Recovery of Bitumen Fractions in the Presence and Absence of Sediment

	Recovery of Fractions from 50 mg Bitumen (Mean of 3 Replicates ± Standard Deviation)	Recovery of Fractions from 50 mg Bitumen Plus 22 g Sediment (Mean of 3 Replicates ± Standard Deviation)	Percent Recovery of Fractions from Sediment Based on Sediment-Free Bitumen Analysis (Mean of 3 Replicates ± Standard Deviation)
Total organic extract, mg	50.4 ± 0.2	50.8 ± 3.2	100.8 ± 6.3
Asphaltene-free extract, mg	37.2 ± 0.4	38.3 ± 1.5	103.0 ± 4.0
Benzene soluble aslphaltenes, mg	9.0 ± 1.0	7.8 ± 0.4	86.7 ± 4.4
Saturates, mg	14.5 ± 1.2	15.2 ± 0.9	104.8 ± 6.2
Aromatics, mg	19.0 ± 0.2	17.7 ± 1.3	93.2 ± 6.8
Polar (NSO) fraction, mg	3.8 ± 0.3	3.7 ± 0.6	97.3 ± 15.8

Since the analyses of dredge samples were performed in triplicate, a means of studying the precision of the method on field samples was available. Combined subsampling and methodological precision for fractionation of triplicate subsamples of dredge sediments, expressed as percent relative standard deviation, is presented in Table 2.2 for two selected concentration ranges. These data not only represent the reproducibility of the fractionation procedure, but also include errors introduced during the subsampling procedure. The general increase in relative standard deviation as the sediment organic content increases is attributed to higher subsampling errors.

Table 2.2   Variation Associated with Separation of Sediment Organic Fractions[a]

Fraction	Concentration Range	
	Low	High
Total Organic Extract		
Range, %	0.25-1.0	1.0-3.0
Number of dredge samples	10	5
Average % relative standard deviation	9.9	27.4
Asphaltene-Free Extract		
Range, μg/g	1300-5000	5000-17750
Number of dredge samples	11	4
Average % relative standard deviation	8.3	12.8
Saturate Fraction		
Range, μg/g	250-500	500-3700
Number of dredge samples	10	5
Average % relative standard deviation	7.2	18.5
Aromatics Fraction		
Range, μg/g	430-1000	1000-6600
Number of dredge samples	8	7
Average % relative standard deviation	5.6	16.8
Polar (NSO) Fraction		
Range, μg/g	420-1000	1000-4900
Number of dredge samples	7	8
Average % relative standard deviation	15.9	17.5
Benzene-Soluble Asphaltenes		
Range, μg/g	450-1000	1000-5400
Number of dredge samples	5	10
Average % relative standard deviation	11.6	14.5

[a]Fifteen sediment dredge samples analyzed in triplicate.

As might be anticipated, the errors associated with field sampling are greater than those associated with subsampling and analysis. Table 2.3 lists the values determined for asphaltene-free extracts from duplicate core samples. The duplicates were taken as closely together as conditions of wind and current would allow. Only one pair of duplicate samples agree within a factor of three for the set of seven duplicates. The variation experienced with other sediment fractions was comparable.

Table 2.3   Analysis of Replicate Core Samples for Asphaltene-Free
Extractable Organic (recoveries expressed as $\mu g/g$ dry sediment)

Location Designation	A	B	C	D	E	F	G
Replicate 1	11,907	9,099	548	<50	1,585	429	220
Replicate 2	16,468	2,041	84	1,536	6,112	1,723	1,292

Relationship of Sediment Organic Content to Sample Location

Because of the high degree of variability in sediment samples, a statistical approach was used to define the relationships between sediment composition and sampling location. Since most intense oil production activity ranges along the northeast shore from Cabimas south to Lagunillas (Figure 2.1), the Battelle Laboratory at Las Morochas was selected as a reference point from which to relate properties of sediment organic fractions taken from other locations. Figures 2.3 and 2.4 show the concentrations of asphaltene-free extract and benzene-soluble asphaltenes as a function of relative distance from the Las Morochas Laboratory. Although the data show extreme variability, a significant correlation ($p < 0.01$) exists between these gravimetric variables.[6] The correlation coefficients of –0.558 and –0.534 reflect the wide variability and indicate that only about 30% of the variability can be accounted for by distance alone. Sampling error and the rather arbitrary selection of the initial point of reference contributed to the variability.

Although the percent saturates in the asphaltene-free extracts were found to be significantly higher in the oil production areas ($p < 0.01$), the correlation coefficient was low (–0.509). Other gravimetric variables investigated did not correlate significantly with distance from the production zone.

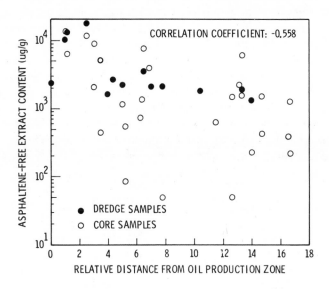

Figure 2.3   Asphaltene-free organic extract *vs.* distance from production zone.

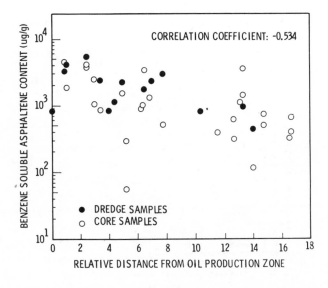

Figure 2.4   Benzene-soluble asphaltenes *vs.* distance from production zone.

Molecular weights and elemental compositions of the organic fractions of dredge samples were correlated with distance from the oil-producing region in a manner similar to the gravimetric analyses. These correlations are illustrated in Figures 2.5-2.9. All variables correlate significantly ($p < 0.05$) with distance from the production region. The correlations of sulfur content and of the molecular weight of the aromatics fractions were also significant at the 0.01 level. There was no significant correlation found between the determined nitrogen contents of any of the fractions and geographical location. Carbon and hydrogen values were consistent with the expected chemical composition of the fractions; *i.e.,* the ratio of hydrogen atoms to carbon atoms in the saturate and aromatic fractions averaged well above 1.8 and below 1.5, respectively.

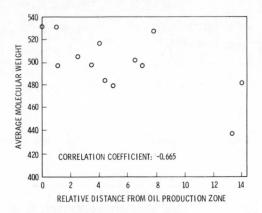

Figure 2.5   Average molecular weight of saturate fraction *vs.* distance from production zone.

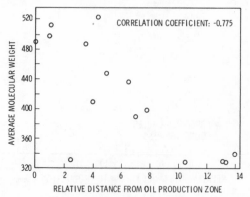

Figure 2.6   Average molecular weight of aromatic fraction *vs.* distance from production zone.

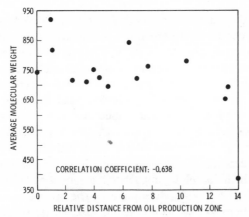

Figure 2.7   Average molecular weight of polar (NSO) fraction *vs.* distance from production zone.

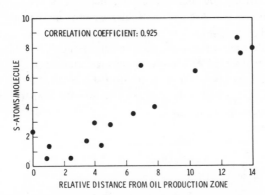

Figure 2.8   Atoms of sulfur per average aromatic molecule *vs.* distance from production zone.

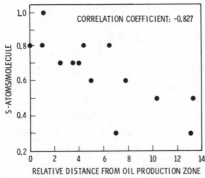

Figure 2.9   Atoms of sulfur per average polar (NSO) molecule *vs.* distance from production zone.

These data show a number of significant trends despite the rather large errors associated with field sampling. The region located in the area of most active oil production is distinctly different in a number of sediment properties. The quantity of organic material extracted as represented by the benzene-soluble asphaltenes and asphaltene-free fractions, the average molecular weights of saturate, aromatic and polar fractions, all decrease as the distance from the production zone increases. The relative quantities of sulfur increase in the aromatics fraction and decrease in the NSO fraction with distance from the oil-producing zone.

The observed trends in sediment composition are consistent with an increase in both the absolute and relative amounts of bituminous residue in sediments from the oil-producing area. The average molecular weights of organic fractions known to be present in asphaltic residue are generally higher than might be expected from similar fractions obtained from recent biological detritus. The presence of this material in sediments would account for the higher average molecular weights in the oil-producing area. A smaller number of sulfur atoms per molecule in aromatic fractions obtained from sediments near the oil-producing zone may be attributed to a dilution of elemental sulfur, found to be present in all aromatics fractions investigated, with aromatic hydrocarbons from bitumen. The presence of hydrocarbon residue in sediments from the producing zone may also bring about losses of sulfur through increased production of hydrogen sulfide.

Determination of Compound-Type Distributions in Sediment Fractions

Although the data indicate that offshore oil production can have a significant effect on the composition of sediments, more specific evidence for the presence of bituminous residues in the sediments of the production zone was obtained by analysis of saturate and aromatic fractions for hydrocarbon types using a high voltage mass spectrometry. The method employed was applied previously by Brown, et al.[7] to characterize hydrocarbons extracted from environmental samples. Using the method of Hood and O'Neal,[8] saturate hydrocarbons were identified and characterized according to number of naphthenic rings from high-voltage cracking patterns. Aromatic compound type distributions were calculated using either the method of Hastings, et al.[9] or of Fitzgerald, et al.[10]

Of the aromatic fractions investigated, 10 out of 25 were found to contain major amounts of sulfur as indicated by the prominent m/e series 258, 256; 226, 224; 192, 194; etc., corresponding to S_8, S_7, S_6, etc. Isotope ratios were consistent with those for amorphous sulfur. None of the samples containing relatively large quantities of sulfur were taken from the oil-producing region. However, sulfur was detected in all aromatics samples. This interfered with the analysis, particularly when the Fitzgerald calculation was used. Using the Hastings method, calculated aromatic compound-type distributions for aromatic fractions containing lesser amounts of sulfur were consistent with that determined for the aromatic fraction of a crude oil obtained from the production zone.

Saturate compound-type distributions determined for sediment samples were significantly different from those obtained for a weathered sample of crude oil from the production area. Figure 2.10 shows the saturate type distributions in sediments from two locations in the producing zone (Stations 2 and 13) and from two locations remote from the oil field

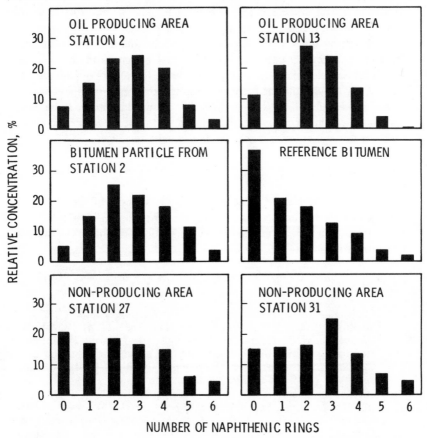

Figure 2.10 Saturate compound-type distribution of sediment organic fractions and reference bitumen.

(Stations 27 and 31). These may be compared with similar measurements on the weathered reference bitumen, and on bituminous particles removed from a dried sediment sample taken from Station 2. The compound-type distribution exhibited by the latter sample is strikingly similar to the saturate extracts from the sediment samples taken from the production zone. This pattern, where the relative abundance of 2- and 3-ring naphthenes predominate, appeared in 14 out of the 25 samples investigated. All but one of these were from the production zone.

It is likely that the relative deficiency of paraffinic hydrocarbons in significant quantities in the saturate fractions of most of the sediment samples examined is a result of physicochemical weathering processes and selective degradation of these compounds, which are more susceptible to microbial decomposition than either naphthenic or aromatic compounds.[11] Thus, the use of crude oils as a reference for identification of bituminous materials in sediments may be of questionable value.

Although bituminous particles are frequently found in the sediments from the petroleum-producing area studied, it is significant that of the 15 dredge samples examined, only three were found to contain the particles, even though the majority of samples exhibited patterns consistent with the presence of bituminous material. It would appear, therefore, that much of the organic material derived from petroleum in the sediments is not present in particles in the visual size range.

ACKNOWLEDGMENTS

The authors are pleased to acknowledge the support of the Creole Petroleum Corporation, Caracas, Venezuela. We also wish to thank Mr. R. L. Buschbom for his assistance and suggestions regarding the statistical treatment of the data.

REFERENCES

1. Clark, R. C. and M. Blumer. *Limnol. Oceanog.* 12, 79 (1967).
2. Mark, H. B. and T. C. Yu. *Environ. Sci. Technol.* 6, 833 (1972).
3. Hunter, L., H. E. Guard and L. H. DiSalvo. "Marine Pollution Monitoring (Petroleum)," Proceedings of a Symposium and Workshop held at NBS, Gaithersburg, Maryland, May 13-17, 1974, *NBS Spec. Publ. 409* (December 1974), p. 213.
4. Sutton, E. A. and R. M. Bean. *Study of Effects of Oil Discharges and Domestic and Industrial Wastewaters on the Fisheries of Lake Maracaibo, Venezuela, Volume I—Ecological Characterization and Domestic and Industrial Wastes,* W. L. Templeton, Ed. (Richland, Washington: Battelle, Pacific Northwest Laboratories, 1974).

5. Koons, B. and P. H. Monaghan. *Data and Discussion of Analyses of the Challenger Knoll Oil*, Initial Report of the Deep Sea Drilling Project, Vol. 1, 478 (1969).

6. Dixon, W. J. and F. J. Massey. *An Introduction to Statistical Analysis*. (New York: McGraw-Hill Book Co., 1951).

7. Brown, R. A., T. D. Searl, J. J. Elliott, B. G. Phillips, D. E Brandon and P. H. Monaghan. *Proceedings of Joint Conference on Prevention and Control of Oil Spills, March 13-15, 1973* (Washington, D.C.: American Petroleum Institute), p. 505.

8. Hood, A. and M. J. O'Neal. *Status of Application of Mass Spectrometry* (New York: Pergamon Press, 1959).

9. Hastings, S. H., B. H. Johnson and H. E. Lumpkin. *Anal. Chem.* **23**, 1243 (1956).

10. Fitzgerald, M. E., V. A. Cirillo and F. J. Galbraith. *Anal. Chem.* **34**, 1931 (1962).

11. ZoBell, C. E. *Proceedings of Joint Conference on Prevention and Control of Oil Spills, December 1969* (Washington, D.C.: American Petroleum Institute), p. 317.

3

THERMAL ALTERATION EXPERIMENTS
ON ORGANIC MATTER IN RECENT MARINE SEDIMENTS
AS A MODEL FOR PETROLEUM GENESIS

M. J. Baedecker,* R. Ikan,*
R. Ishiwatari* and I. R. Kaplan
Institute of Geophysics and Planetary Physics
University of California
Los Angeles, California 90024

INTRODUCTION

Studies on the thermal alteration of organic matter provide information
on degradative pathways that occur during diagenesis. By conducting
these experiments in the laboratory, it is possible to control the conditions of degradation and .observe changes that are difficult or impossible
to monitor in the natural environment. Few studies have been reported
on the heat treatment of unlithified sediments.[1-3] This study was
undertaken to determine the fate of naturally occurring lipids and pigments in a marine sediment exposed to elevated temperatures. This was
accomplished by heating samples of a young marine sediment from
Tanner Basin to a series of temperatures ($65°$-$200°C$) for varying periods
of time (7-64 days). The sediment was analyzed prior to and after
heating for pigments, isoprenoid compounds and alcohols (for which data
have been published[4-6]), and fatty acids and hydrocarbons. This chapter

*Present Addresses: Baedecker–U.S. Geological Survey, Mail Stop 432, Reston,
Virginia 22092; Ikan–Organic Chemistry Department, Natural Products Laboratory,
Hebrew University, Jerusalem; Ishiwatari–Department of Chemistry, Faculty of
Science, Tokyo Metropolitan University, Setagaya-Ku, Tokyo, Japan.

is devoted to a summary of results on isoprenoids and pigments containing the tetrapyrrole structure and provides additional data for normal alkanes and fatty acids. Studies on the unextractable organic matter or kerogen were also undertaken to determine the structural changes that occur upon heating. The results are intended to furnish insight into mechanisms of converting precursors to products as related to the genesis of petroleum.

METHODS

Fifty-gram samples of sediment from Tanner Basin were sealed, under nitrogen, in thick-walled glass tubes and subjected to heat treatment at temperatures of 65°, 100°, 150° and 200°C for varying periods of time (7, 30 and 64 days).

Extractable organic material was removed from the sediment with benzene-methanol according to the method of Brown, *et al.*[7] A crude separation of the extract was achieved by column chromatography on silicic acid using the following eluants: hexane, benzene and methanol. Normal and branched hydrocarbons, which were in the hexane fraction, were separated by 5 Å molecular sieves. A portion of the benzene and methanol fractions (from the silicic acid column) were combined for analysis of fatty acids and alcohols (see references 6 and 7, respectively, for more details). After methylation with diazomethane, the branched and normal fatty acid methyl esters were separated by urea adduction.[8] The fractions were purified by thin layer chromatography on silica gel F-254 plates (0.3 mm) using as solvent system petroleum ether-ether-acetic acid (80:20:1).

Identification of individual hydrocarbons and fatty acids was made by gas liquid chromatography using a 5 ft x 1/8 in. column packed with 3% OV-101 on 100-120 mesh Gas Chrom Q, programmed from 100° to 300°C at 4°/minute. Mass spectra were obtained using a Varian Model 1200 gas chromatograph coupled with a 21-491 CEC mass spectrometer. Compound identification was obtained from the retention times and mass spectra by comparison with those of known standards. The reference compounds were available commercially with the exception of the C_{18}-ketone and the olefinic isoprenoid hydrocarbons, which were synthesized.

Details for the separation of tetrapyrrole pigments and their characterization by thin layer chromatography and ultraviolet and mass spectroscopy are published elsewhere.[4]

Kerogen was isolated from the sediment by the method developed by Saxby[9] with the exception that the nitric acid treatment was omitted. Carbon and hydrogen were measured in KBr pellets using a Perkin-Elmer

421 spectrometer. Electron spin resonance measurements were performed with thoroughly dried, finely ground samples in standard ESR quartz tubes (3 mm internal diameter), filled to a height of 3 cm. The weight of each sample in the tube was determined by weighing first the empty tube and then the tube plus sample. ESR spectra were recorded on a Varian Associates E-12 spectrometer, employing 100 KHz modulation and a normal operating frequency of 9.49 GHz. The magnetic field (g) of the samples was calibrated with DPPH (diphenyl-picrylhydrazyl) supplied by Varian Associates. Spin concentrations were estimated by comparison with a standard strong pitch in powdered KCl (3×10^{15} spins/cm). The number of free radicals was assumed to be proportional to signal height times signal width squared. The peak-to-peak separation of the signal was taken as line width. Spectroscopic splitting factors (g-values) of samples were approximated from values of the magnetic field at which resonance occurred relative to standards of known g-values (DPPH, g = 2.0036; standard pitch, g = 2.0028).

RESULTS AND DISCUSSION

Tetrapyrrole Pigments

There are two major types of tetrapyrrole pigments: chlorins (Figure 3.1), which are generally found in recent sediments,[10,11] and porphyrins, which are associated with ancient sediments.[12-14] Characterization of

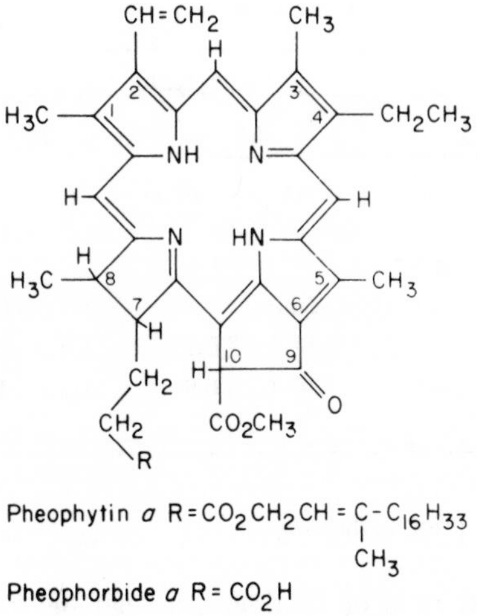
CH=CH2
CH3
H
H3C
CH2CH3
NH
N
H
H
N
HN
H
H3C
CH3
H
CH2
H
CH2
CO2CH3
O
R
Pheophytin a R=CO2CH2CH=C-C16H33
CH3
Pheophorbide a R=CO2H

Figure 3.1 Structures of pheophytin α and pheophorbide α.

these compounds has been difficult because of the complex mixtures involved. However, the recent application of mass spectrometry to pigment studies has provided the means to elucidate their structures.[15,16] Chlorins in the marine environment are derived mainly from chlorophyll, although chlorophyll itself has seldom been reported in recent sediments.[17,18] Pheophytin *a* and *b* (loss of Mg from the chlorophyll molecule) and pheophorbide *a* and *b* (loss of Mg and the phytyl group) are the chlorins that have been identified most frequently. Porphyrins found in petroleum bear a close resemblance to the chlorophyll structure. However, if derived from chlorins they have undergone several transformations, including dehydrogenation at the 7,8 position, reduction of the 2-vinyl group and 9-keto group, decarboxylation and, in some cases, cleavage of the isocyclic ring. Petroleum porphyrins are highly alkylated and usually associated with metals, especially Ni and V. Few compounds have been reported in sediments that are intermediate between pheophytins—pheophorbides and porphyrins; Hodgson[19] has reported a possible Ni complex of pheophytin, and Baker and Smith[20] have reported a series of chlorins that are characterized by the opening of the isocyclic ring and simultaneous hydrolysis of the phytol ester.

Our studies showed a 20-fold loss of chlorins in sediment heated to 100°C and no detectable chlorins in sediment heated to 150°C (Table 3.1). Temperature appears to have a greater influence on the destruction

Table 3.1   Chlorins (C) and Porphyrins (P) from Tanner Basin Sediment (concentration in ppm)

Time of Heating (days)	Temperature (°C)							
	25°		65°		100°		150°	
	C	P	C	P	C	P	C	P
0	20.7	n.d.[a]						
7			26.0	0.8	1.7	0.6	n.d.	1.0
30			17.9	0.9	1.6	1.5	n.d.	2.0
64	34.0	n.d.	23.4	0.9	1.1	1.5	n.d.	6.7
64								3.3[b]

[a]n.d. = not detected
[b]Sediment freeze-dried prior to heating.

of chlorins than does time of heating, although the longest heating time in these experiments was only 64 days, which may be too short for any significant geological comparison. Of particular interest is the formation of porphyrins (not present initially in the sediment) at concentration levels as high as 6 ppm in sediment heated to 150°C for 64 days. The small conversion (maximum of 20-30%) of chlorins to porphyrins may be due to several factors, such as destruction of the chlorins, polymerization (which would yield high-molecular-weight compounds difficult to extract from the sediment) and incorporation into kerogen (also nonextractable).

The chlorins identified were predominantly of the pheophytin and pheophorbide type, with larger amounts of the *a* series than *b* series. Interestingly, compounds with saturated side chains, namely the 2-vinyl group and the double bond of the phytyl group in pheophytins (Figure 3.1), were dominant. Thus, in a reducing environment, the phytyl double bond is hydrogenated early in diagenesis which, as suggested by Smith and Baker,[11] inhibits loss of the phytyl group by an "internal elimination." This may explain the extremely low levels of isoprenoid hydrocarbons found in recent sediments as compared to ancient sediments. Small amounts of chlorins characterized by opening of the isocyclic ring and cleavage of the phytyl group (chlorin e_4, e_6 types) were tentatively identified. These compounds are probably precursors to porphyrins of the etio type. The porphyrins that formed upon heating were a complex mixture. Spectral evidence indicated that some of the porphyrins were complexed with nickel or vanadium.

Isoprenoid Compounds

Since the discovery of isoprenoid hydrocarbons in petroleum,[21,22] a great deal of interest has developed in searching for the precursors of these compounds. This has led to the identification of isoprenoidal hydrocarbons, alcohols, ketones and acids in sediments and in living systems. Alkanes have been found in petroleum and oil shales with carbon chain lengths ranging from 14-24.[23,24] However, those most frequently reported are pristane (2,6,10,14-tetramethylpentadecane) and phytane (2,6,10,14-tetramethylhexadecane). Bendoraitis[22] suggested that pristane originated from chlorophyll and, although this theory was extended by others,[25] the intermediate compounds have not been isolated, at least not within any one sediment.

We were able to identify a series of isoprenoid compounds from sediment heated at 100° and 150°C for 30 days. They are listed along with their concentrations in Figure 3.2. With the exception of the C_{18} ketone none of these compounds were present initially in the sediment. In Figure 3.3 we have outlined the most feasible pathway of chlorophyll degradation, which accounts for the compounds identified in this study.

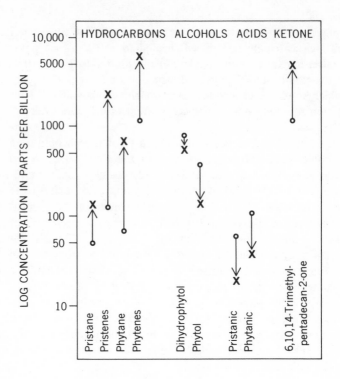

Table 3.2 Concentrations of isoprenoid compounds found in Tanner Basin sediment heated for 30 days (0 = 100°C; X = 150°C).

The formation of the C_{18} ketone requires an oxidation step, and on the basis of its large concentration in a mildly reducing sediment it was probably formed prior to sedimentation and released during heating. Its fate during diagenesis is somewhat uncertain; the expected pathway would be hydrogenation and subsequent dehydration to yield C_{18} hydrocarbons. An alternative pathway could be the direct reduction of the ketone to the corresponding alkane. Results published elsewhere[26] show the ketone concentration decreases substantially in sediment heated at 150°C for 64 days, yet the C_{18} alcohol and alkene were not found in the sediment and only traces of the alkane were detected.

As shown in Figure 3.2, the hydrocarbons were more abundant at the higher temperature (150°C) than alcohols and acids. This suggests a conversion of the more polar compounds to the nonpolar and more stable hydrocarbons. Since the abundance of the isoprenoid hydrocarbons was substantially above their detection limit, we were able to determine their concentrations in samples heated at 100°C and 150°C for various

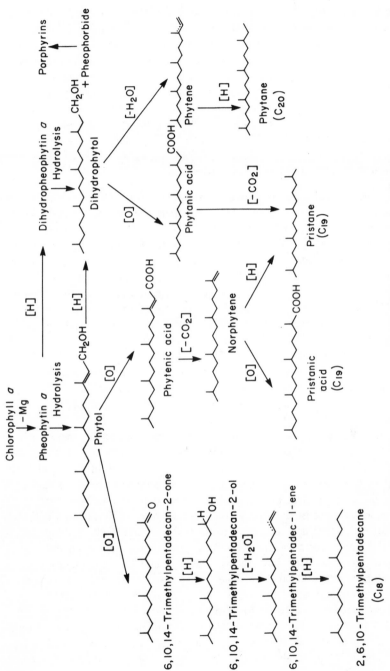

Figure 3.3   Proposed pathway for chlorophyll degradation in sediment.

periods of time. No hydrocarbons were detected in the unheated sediment or in sediment heated at 65°C. As seen from Table 3.2, there is a direct correlation between time of heating and the amount of hydrocarbons formed. It is of interest that phytane is formed before pristane during heating and is also more abundant. Thus, the degradation of chlorophyll in a generally reducing environment would favor the formation of phytane via dihydrophytol; however, exposure to an oxidizing environment during early diagenesis would favor the formation of phytenic acids, which would ultimately yield pristane.

Normal Hydrocarbons and Fatty Acids

Normal hydrocarbons represent a small portion of the extractable compounds in recent sediments. However, their distribution reflects the source of material that is deposited, in that a greater abundance of hydrocarbons in the C_{21} to C_{31} range with a high Carbon Preference Index (CPI) (ratio of hydrocarbons with odd carbon numbers to hydrocarbons with even carbon numbers) indicates terrestrial input while a greater abundance in the C_{15} to C_{22} range indicates contribution from marine organisms. In many environments, a bimodal distribution is observed where there are near-equal contributions of marine and terrestrial organic matter.[7] During lithification and subsequent alteration of organic matter, the normal hydrocarbon curve (C_{12} to C_{34}) undergoes a smoothing effect as the CPIs of the long-chained components are decreased and lower-molecular-weight components are formed. In petroleum, the CPI of normal alkanes decreases until in Mesozoic and Paleozoic deposits, values of ~ 1 are reached.[28]

Our studies on hydrocarbons extracted from heated sediment, although restricted to a few samples, clearly show that changes are occurring during maturation. Figure 3.4 shows the gas chromatograms of the total hydrocarbons extracted from unheated sediment, sediment heated at 150°C for 64 days and freeze-dried sediment heated to 150°C for 64 days. The major peak in the natural sediment (Figure 3.4B) was not definitely identified. It is a labile constituent, as it was essentially destroyed during the sieving process to separate normal from branched components. Youngblood, et al.[29] have shown that C_{19} and C_{21} polyolefins are abundant in benthic marine algae. They also determined that olefins with a terminal double bond in the 3- position are characterized by a strong mass/ion charge (m/e) of 55 and by the loss of an ethyl group (M-29). The mass spectrum of the compound in question showed significant peaks at m/e 259 and 55, which would be consistent with a C_{21}-tetraolefin of molecular weight 288. Figure 3.4A shows that this compound

Table 3.2 Concentration (ppm) of Isoprenoid Hydrocarbons in Heat-Treated Tanner Basin Sediments

	C_{18}-Alkane[a]	Pristane	Pristene-1[b]	Pristene-2[c]	Phytane	Phytene-1[d]	Phytene-2[e]
100°C/7 days	n.d.	n.d.	0.07	tr.	0.06	0.25	0.37
100°C/30 days		tr.	0.13	tr.	0.07	0.44	0.78
100°C/64 days		tr.	0.40	tr.	0.17	0.92	1.45
150°C/7 days		0.07	1.04	0.34	0.40	1.84	2.55
150°C/30 days		0.13	1.39	0.82	0.64	2.57	3.43
150°C/64 days	0.08	0.65	0.68	3.30	1.98	3.98	3.38

[a]2,6,10-trimethylpentadencane
[b]2,6,10,14-tetramethylpentadec-1-ene
[c]2,6,10,14-tetramethylpentadec-2-ene; tentatively identified on basis of GLC-MS data. The monolefins, pristene-1 and pristene-2, are usually referred to as norphytenes.
[d]3,7,11,15-tetramethylhexadec-1-ene
[e]3,7,11,15-tetramethylhexadec-2-ene

NOTE: The alkene isomers had the same mass spectra; therefore assignment of the double bond was made by comparison with the results of Blumer and Thomas[27] who showed the -1-ene has a lower retention time than the -2-ene.

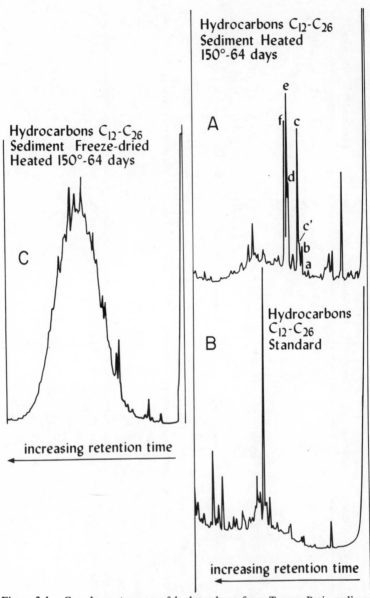

Figure 3.4 Gas chromatograms of hydrocarbons from Tanner Basin sediment in **A.** Sediment heated at 150°C for 64 days, **B.** Unheated sediment, and **C.** Freeze-dried sediment heated at 150°C for 64 days. Labeled peaks are as follows: a - C_{18} isoprenoid alkane; b - pristane; c' - pristene-1; c - pristene-2; d - phytane; e - phytene-1 and f - phytene-2 (*see* Table 3.2).

is destroyed by heat treating the sediment. When the sediment was freeze-dried prior to heat treatment (Figure 3.4C), a hydrocarbon hump appears that could not be resolved by gas chromatography on a 50-foot capillary column.

Changes also occurred in the abundance of normal hydrocarbons which, due to the small amounts present relative to the total hydrocarbons, were not recognizable without sieving. A plot of the normal hydrocarbon content *vs.* length of the carbon chain for these same three samples (Figure 3.5) shows a rearrangement of the n-alkanes during heating, resulting in the decrease of long chain compounds and increase of the lower-molecular-weight components. This trend appears to be accentuated by freeze-drying

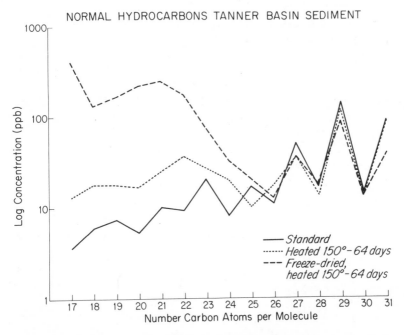

Figure 3.5 Concentration of normal hydrocarbons (C_{17}-C_{31}) of natural and heat-treated ($150°C$ for 64 days) Tanner Basin sediment.

prior to heat treatment. Further, the CPI decreased (Table 3.3) from 5.3 to 1.4 and, as before, the freeze-dried heated sample showed more alteration (lower CPI) of hydrocarbons than those produced by heating the wet sample.

Freeze-drying the sediment prior to heating removes coordinated water from clay mineral surfaces, and thus they probably act as more efficient

Table 3.3 Hydrocarbons (HC) and Fatty Acids (FA) in Tanner Basin Sediment

Time of Heating (days)	Sediment	Temp. (°C)	Total HC (ppm)	n-HC ppm	n-HC CPI[a]	i-HC[b] (ppm)	n-FA (ppm)
0	wet	25	38.8	0.4	5.3	n.d.	1.7
64	wet	150	58.4	0.5	2.8	14.0	21.4
64	freeze-dried	150	308.0	1.7	1.4	n.d.[c]	19.0

$$
{}^{a}\mathrm{CPI} = \frac{2 \, \Sigma \text{ odd } n\text{-}C_{21} \text{ to } n\text{-}C_{31}}{\Sigma \text{ even } n\text{-}C_{20} \text{ to } n\text{-}C_{30} + \Sigma \text{ even } n\text{-}C_{22} \text{ to } n\text{-}C_{32}}
$$

[b]Isoprenoid hydrocarbons.

[c]Isoprenoids may be present in this sample, but obscured by a "hydrocarbon hump" in the C_{16} to C_{24} region (Figure 3.4C).

Lewis acids. The catalytic cracking of alkanes can proceed by a carbonium ion mechanism, involving an acid catalyst, which provides the protons for interacting with the hydrocarbons.[30,31] We found that heating dried sediment for short periods of time (1-7 days) at relatively low temperatures (200°C) produced up to a 10-fold increase of hydrocarbons (Figure 3.6), relative to the amount present in the natural sediment.

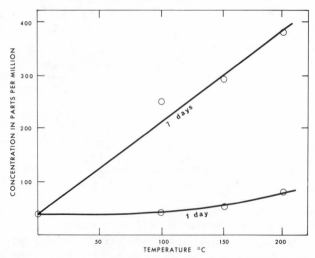

Figure 3.6 Concentration (by weight) of the total hydrocarbons in natural sediment (40 ppm) and in freeze-dried sediment heated at 100°, 150° and 200°C for 1 day and 7 days.

The reaction mechanisms for hydrocarbon production are complex. In experiments by Hoering,[3] where dried sediment previously exposed to deuterium oxide was heated to 225°C, normal hydrocarbons formed which were deuterated at several sites. More complex reactions are involved than simply a release of alkyl groups from polymeric structures. Reduction of organic compounds on mineral surfaces in the presence of deuterated species could lead to deuterium replacement at positions other than the reaction site,[3] providing an additional role for minerals in catalyzing organic reactions in the natural environment.

The summary of results in Table 3.3 also gives some data for normal fatty acids. Although the fatty acids increased in abundance in the heated sediment, there appeared to be little difference whether the sediment was wet or dried prior to heating. The distribution of normal fatty acids was not changed: C_{16} was always the dominant acid. More branched acids were formed during heating relative to the normals; however, no attempt was made to identify specific compounds, with the exception of the isoprenoidal acids. It appears that the heating of sediment at low temperature for short periods of time either causes release of the fatty acids from sediment surfaces, or cleaves them off complexes such as fulvic or humic acids.

Kerogen Studies

Kerogen was isolated from sediment to observe possible structural changes that may occur during heating of the sediment. There is a progressive increase in kerogen with increasing temperature and time of heating. The kerogen content and percentages of carbon and hydrogen are given in Table 3.4. The value for kerogen in the unheated sample (4.2%) is surprisingly low; however, the same value was obtained during a repeated analysis. Part of this increase may be due to a thermal grafting of monomers, which are formed from the breakdown of biological polymers during heating. There is evidence of change in the polymeric network of kerogen in that the carbon-to-hydrogen ratios are increased in proportion to time and temperature of heating (Table 3.4).

It has been recognized for some time that polymers such as lignins, humic acids and coals contain free radicals. The nature of these long-lived free radicals is not well defined. Steelink[32] provided evidence that the free radicals in lignin and humic acids from soil and lignite are probably semiquinone structures coexisting with quinhydrones that have been stabilized by shielding of the polymeric network. In asphaltenes, which have a low (< 4%) oxygen content, the free electrons may be

Table 3.4 Data for Kerogen Isolated from Heat-Treated Sediment

Sediment	% Total Organic Matter in Sediment[a]	% Kerogen in Sediment	(Humic Acid + Fulvic Acid) Kerogen	%C in Kerogen	%H in Kerogen	C/H	Spins per g-Kerogen x 10^17	g-Value
Unheated	9.5	4.2	1.0	56.0	6.6	8.5	1.0	2.0032
65° - 7 days	9.4	7.0	0.2	57.1	6.9	8.4	1.6	2.0030
65° - 30 days	9.7	8.5	0.1	57.7	6.6	8.8	1.6	2.0030
100° - 7 days	8.6	7.5	0.01	60.2	6.1	9.9	2.3	2.0030
100° - 30 days	8.9	9.0	0.01	59.4	6.5	9.1	2.2	2.0028
150° - 7 days	8.1	7.7	0.01	60.8	6.6	9.2	2.3	2.0030
150° - 30 days	9.4	9.4	0.03	58.3	6.0	9.7	2.4	2.0030

[a]Total Organic Matter = Total Organic Carbon x 1.8

associated with the aromatic system and stabilized by delocalization.[33] While the kerogen we isolated was not sufficiently modified during heating to structurally approximate asphaltene, certain trends of alteration can be observed. In Tanner Basin humic material, the combined content of O, S, N (by difference of C and H determinations) was 41.6%[34] and for kerogen the values ranged from 37.4% for unheated sediment to 33.2% for sediment heated at 150°C. The observed loss of polar constituents and increase in aromaticity apparently begins in the early stages of diagenesis and proceeds systematically with increasing temperature. This was confirmed by analysis of the infrared spectra, which showed less absorbance at 2800-3000 cm^{-1} (due to C-H alkane stretching) and at 1700 cm^{-1} (due to C=O stretching) in the kerogens from heat-treated samples.

The concentrations of stable free radicals can be determined by electron spin resonance (ESR). ESR studies of pure kerogens were initiated by Marchand et al.[35] and the technique was extended by Pusey[36] as a method for evaluating paleotemperatures and thus potential maturation of source rocks.

ESR measurements on kerogens show an increase in spin concentrations from 1.0×10^{17} in unheated sediment to 2.4×10^{17} in sediment heated to 150°C for 30 days (Table 3.4). The ESR spectra of kerogens measured at 100 G consist of a single symmetrical line devoid of any splitting, which indicates that the free radicals are extremely complex in structure. Spin concentrations of complex material obtained in one study are difficult, if not impossible, to compare with analyses made by others, as the determination is extremely sensitive to the methods used to prepare samples, standards used for calibration and the geometry of the system.

The g-values, which are independent of the amount of organic matter present, did not show any marked change during heating of the sediment from an initial value of 2.0032. This value compares closely with those obtained by Yen, et al.[33] of 2.0031 ± 0.0005 for petroleum asphaltenes. The g-values of coals and kerogens decrease during maturation,[36,37] which is mainly associated with heat generation during burial.

Alterations in the structure of kerogen during burial are important in evaluating the petroleum potential of sediments.[38] Kerogens from marine environments such as Tanner Basin, which contains lower carbon-to-hydrogen ratios, should be better producers of oil. Tissot et al.[39] simulated kerogen maturation by heating samples to 600°C and observed the release of aliphatic groups and loss of oxygen-containing groups. By heating sediment at lower temperatures for longer periods of time, we observed similar transformations, but to a lesser extent. In our experiments the overall processes involve formation and transformation of kerogen rather than degradation.

The overall pathway of alteration at the low temperatures at which the experiment was performed appears to be complexing of lower-molecular-weight compounds and then loss of oxygen-rich compounds (probably as CO_2 and H_2O). Confirmation of this comes from virtual loss of humic acid in the sediment at temperatures of $100°C$ and higher (Table 3.4) and almost quantitative conversion of all organic matter to kerogen within 30 days. At the same time, hydrogen is preferentially lost in comparison to carbon.

CONCLUSIONS

The mildly reducing Tanner Basin sediment has a high organic carbon content (> 5%) and conditions favorable for the preservation of organic matter. Our experiments on the thermal alteration of organic matter in sediments indicated that the following processes occur during early maturation. Fatty acids and hydrocarbons increased in abundance although there appeared to be no obvious precursor-to-product relationship via simple decarboxylation reactions. Chlorins were partially converted into porphyrins. The phytyl side chain of pheophytin was initially preserved intact by reduction of the phytyl double bond, but later converted to a variety of isoprenoid compounds including alkanes. Some components were apparently thermally grafted onto kerogen as the concentration of kerogen increased on heating. However, structural changes also were occurring, evidenced by an increase in aromaticity, loss of oxygen-containing groups and changes in the ESR spectra.

ACKNOWLEDGMENTS

We wish to thank Dr. Kai Fang for assistance in determining ESR spectra and Beth Irwin for assistance with mass spectroscopic analysis. This study was supported by a grant from NASA, Office of Exobiology, number NGR 05-007-221.

REFERENCES

1.   Douglas, A. G., G. Eglinton and W. Henderson. *Advances in Organic Geochemistry, 1966* G. D. Hobson and G. C. Speers, Eds. (New York: Pergamon Press, 1970), p. 369.
2.   Eglinton, G. *Advances in Organic Geochemistry, 1971*, H. R. v. Gaertner and H. Wehner, Eds. (New York: Pergamon Press, 1971), p. 29.
3.   Hoering, T. C. *Carnegie Institute Yearbook 67*, 199 (1967).

4. Ikan, R., Z. Aizenshtat, M. J. Baedecker and I. R. Kaplan. *Geochim. Cosmochim. Acta* **39**, 173 (1975).
5. Ikan, R., M. J. Baedecker and I. R. Kaplan. *Geochim. Cosmochim. Acta* **39**, 187 (1975).
6. Ikan, R., M. J. Baedecker and I. R. Kaplan. *Geochim. Cosmochim. Acta* **39**, 195 (1975).
7. Brown, F. S., M. J. Baedecker, A. Nissenbaum and I. R. Kaplan. *Geochim. Cosmochim. Acta* **36**, 1185 (1972).
8. Douglas, A. G., M. Blumer, G. Eglinton and K. Douraghi-Zadeh. *Tetrahedron* **27**, 1071 (1971).
9. Saxby, J. D. *Geochim. Cosmochim. Acta* **34**, 1317 (1970).
10. Orr, W. L., K. O. Emery and J. R. Grady. *Bull. Amer. Assoc. Petrol. Geol.* **42**, 925 (1958).
11. Smith, G. D. and E. W. Baker. *Initial Reports of the Deep Sea Drilling Project*, Volume XXII (Washington, D.C.: U.S. Government Printing Office, 1974), p. 677.
12. Treibs, A. *Angew. Chem.* **49**, 682 (1936).
13. Hodgson, G. W., B. Hitchon, K. Taguchi, B. L. Baker and E. Peake. *Geochim. Cosmochim. Acta* **32**, 737 (1968).
14. Thomas, D. W. and M. Blumer. *Geochim. Cosmochim. Acta* **28**, 1147 (1964).
15. Baker, E. W., T. F. Yen, J. P. Dickie, R. E. Rhodes and L. F. Clark. *J. Amer. Chem. Soc.* **89**, 3631 (1967).
16. Jackson, A. H., G. W. Kenner, K. M. Smith, R. T. Aplin, H. Budzikiewicz and C. Djerassi. *Tetrahedron* **21**, 2913 (1965).
17. Drozdova, T. V. and Y. N. Gurskii. *Geokhimiya No. 3*, 323 (1972); *Geochim. Internat.*, 208 (1972).
18. Nissenbaum, A., M. J. Baedecker and I. R. Kaplan. *Geochim. Cosmochim. Acta* **36**, 709 (1972).
19. Hodgson, G. W., B. Hitchon, R. M. Elofson, B. L. Baker and E. Peake. *Geochim. Cosmochim. Acta* **19**, 272 (1960).
20. Baker, E. W. and G. D. Smith. *Initial Reports of the Deep Sea Drilling Project*, Volume XX (Washington, D.C.: U.S. Government Printing Office, 1973), p. 193.
21. Dean, R. A. and E. V. Whitehead. *Tetrahedron Lett.* **21**, 768 (1961).
22. Bendoraitis, J. G., B. L. Brown and L. S. Hepner. *Anal. Chem.* **34**, 49 (1962).
23. Han, J. and M. Calvin. *Geochim. Cosmochim. Acta* **33**, 733 (1969).
24. Spyckerelle, C., P. Arpino and G. Ourisson. *Tetrahedron* **28**, 5703 (1972).
25. Cox, R. E., J. R. Maxwell, R. G. Ackman and S. N. Hooper. *Advances in Organic Geochemistry, 1971* H. R. v. Gaertner and H. Wehner, Eds. (New York: Pergamon Press, 1972), p. 263.
26. Ikan, R., M. J. Baedecker and I. R. Kaplan. *Nature* **244**, 154 (1973).
27. Blumer, M. and D. W. Thomas. *Science* **148**, 370 (1965).
28. Bray, E. E. and E. D. Evans. *Geochim. Cosmochim. Acta* **22**, 2 (1961).

29. Youngblood, W. W., M. Blumer, R. L. Guillard and F. Fiore. *Mar. Biol.* **8**, 190 (1971).
30. Johns, W. D. and A. Shimoyama. *Bull. Amer. Assoc. Petrol. Geol.* **56**, 2160 (1972).
31. Greensfelder, B. S., H. H. Voge and G. M. Good. *Ind. Eng. Chem.* **41**, 2573 (1949).
32. Steelink, C. *Geochim. Cosmochim. Acta* **28**, 1615 (1964).
33. Yen, T. F., J. G. Erdman and A. J. Saraceno. *Anal. Chem.* **34**, 695 (1962).
34. Nissenbaum, A. and I. R. Kaplan. *Limnol. Oceanog.* **17**, 570 (1972).
35. Marchand, A., P. A. Libert and A. Combaz. *Compt. Rend. Ser. D* **266**, 2316 (1968).
36. Pusey, W. C. *Trans. Gulf Coast Assoc. Geol. Soc.* **23**, 195 (1973).
37. Retcofsky, H. L., J. M. Stark and R. A. Friedel. *Anal. Chem.* **40** 1999 (1968).
38. McIver, R. D. *Proc. 7th World Petrol. Congr., Mexico* **2**, 25 (1967).
39. Tissot, B., J. Durand, J. Espitalie and A. Combaz. *Bull. Amer. Assoc. Petrol. Geol.* **58**, 499 (1974).

4

FOSSIL PORPHYRINS AND CHLORINS
IN DEEP OCEAN SEDIMENTS

Earl W. Baker and G. Dale Smith

Department of Chemistry
Northeast Louisiana University
Monroe, Louisiana 71201

INTRODUCTION

In an earlier paper[1] we described recent advances in separation and analytical methods available for investigation of pigments in marine sediments. Improvements were noted in separation-chromatography techniques and comparisons were made in chemical, spectral and chromatographic characteristics of various examples of chlorin and porphyrin tetrapyrroles. The application of mass spectrometry to the analysis of tetrapyrroles was shown to be a most powerful analytical method. Readable mass spectra were obtained on small labile samples, 5-15 μg of complex pigment mixtures.

The advent of such sophisticated analytical methods to organic geochemistry opened the door to investigations of minute samples of pigment contained in the sediments buried beneath the ocean floor. Thus, the chemical diagenesis of chlorophylls in marine sediments has received increased interest in recent years. Substantial evidence has been accumulated indicating that the chlorophylls are the precursors of metalloporphyrins occurring in a wide variety of asphaltenes and other bituminous materials.[1-8] These data support and indeed almost may be said to confirm Treibs' original hypothesis of the biological origin of petroleum.[2]

In 1934 Treibs defined a "biological marker" as a compound having enough of its carbon skeleton preserved through accumulation and diagenesis

73

to still allow it to be correlated with its original biological precursor. He then demonstrated the power of this concept by relating the metalloporphyrins in bitumens to biological tetrapyrroles and concluded that petroleum was biological in origin. Since then, a self-consistent theory of organic geochemistry has grown from this base.

Considerable knowledge of the structures and properties of the chlorophyll precursors and the thermally stable metalloporphyrin end products has been gained from studies of pigments occurring in oxygen-evolving green photosynthetic plants[9] and various bitumen sources[1-8,10-26] (see Figure 4.1a and 4.1b). However, details of intermediate products and mechanisms of the diagenetic pathways of the conversion of chlorophyll to petroporphyrin are difficult, if not impossible, to ascertain from the presently published data collected from investigations of samples of greatly diverse geoenvironmental history.

In the present study, 32 core samples recovered by the "Glomar Challenger" of the Deep Sea Drilling Project were obtained for analysis of tetrapyrrole pigment content (see Table 4.1). The samples were selected to provide a continuous history of the diagenesis of chlorophyll to petroporphyrins in marine environments. Particular cores were chosen with the assumption that samples near the sediment-water interface contained products of early diagenesis while those at successively greater depths gave rise to pigments having undergone more severe thermal alteration. Thus, the investigation concentrated on a suite of cores expected to provide significant and consistent information on the intermediate products from which mechanisms for the degradation pathways could be deduced for the total diagenesis of chlorophyll.

EXPERIMENTAL PROCEDURE AND EQUIPMENT

Treatment of Cores

A flow sheet outlining the experimental procedures used in treatment of core samples is given in Figure 4.2.

The frozen cores were broken up into approximately 120-g samples and exhaustively extracted with 90% acetone-10% methanol using a 1.0-liter ball mill. The sediment sample was rolled at low speed in the dark at room temperature for 4 hr with approximately 250 ml solvent. The solvent was removed by vacuum filtration and the procedure repeated until all pigment had been extracted.

The combined extracts were reduced to dryness, taken up in freshly distilled tetrahydrofuran (THF) and chromatographed on Sephadex LH-20 (for column preparation, see Baker, 1970). The various pigment-containing

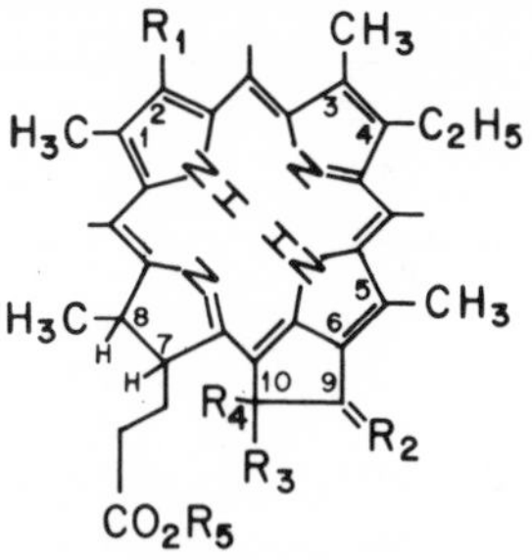

Name	R_1	R_2	R_3	R_4	R_5
Pheophytin a	$CH=CH_2$	O	$COOCH_3$	H	$CH_2CH=\underset{\underset{CH_3}{\vert}}{C}-C_{16}H_{33}$
Pheophorbide a	$CH=CH_2$	O	$COOCH_3$	H	H
Dihydrophytol Pheophorbide a	$CH=CH_2$	O	$COOCH_3$	H	$CH_2CH_2\underset{\underset{CH_3}{\vert}}{CH}-C_{16}H_{33}$
Dihydrophytol-10-hydroxy pheophorbide a	$CH=CH_2$	O	$COOCH_3$	OH	$CH_2CH_2\underset{\underset{CH_3}{\vert}}{CH}-C_{16}H_{33}$

Figure 4.1a Structures of pheophytin analogues.

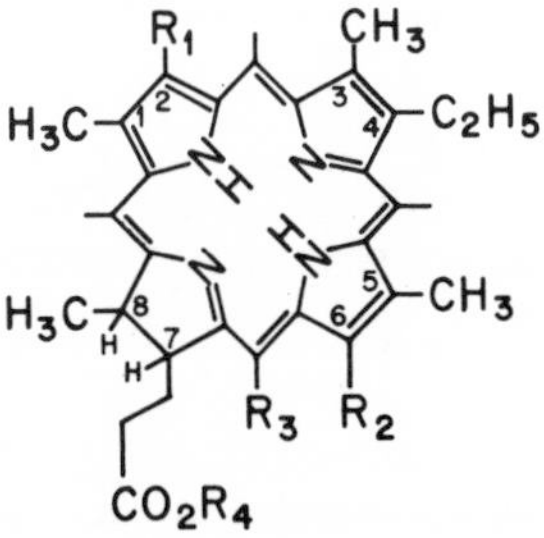

Name	R_1	R_2	R_3	R_4
Dihydrophytol chlorin P_6 DME	$CH=CH_2$	CO_2CH_3	CO_2CH_3	$CH_2CH_2\underset{\underset{CH_3}{\vert}}{CH}-C_{16}H_{33}$
Chlorin P_6 TME	$CH=CH_2$	CO_2CH_3	CO_2CH_3	CH_3
Mesochlorin P_6 TME	CH_2-CH_3	CO_2CH_3	CO_2CH_3	CH_3
Purpurin 18	$CH=CH_2$	$OC\diagdown_O\diagup CO$		H

Figure 4.1b Structures of pheophytin derivatives.

Table 4.1 DSDP Core Description

DSDP Core	Depth[a] (meters)	Age[a]	Pigment Yield[b] (ppb)
14-144A-02-06	45	Eocene	5
14-144B-03-06	35	Oligocene	30
15-147B-01-03	1	Pleistocene	28,000
15-147B-02-01	10	Pleistocene	30,000
15-147B-05-00	40	Pleistocene	29,000
15-147B-05-06	49	Pleistocene	14,000
15-147B-07-04	65	Pleistocene	2,000
15-147B-09-00	78	Pleistocene	2,100
15-147C-02-03	107	Pleistocene	1,800
15-147C-05-02	140	Pleistocene	600
20-198A-04-03	125	U. Cretaceous	b.d.
22-217-02-04	44	Pliocene	10
22-217-10-04	348	Eocene	b.d.
22-217-18-02	432	U. Cretaceous	b.d.
22-217-27-02	517	U. Cretaceous	b.d.
22-217-32-02	564	U. Cretaceous	10
22-218-05-02	71	Pleistocene	80
22-218-11-02	300	Pliocene	b.d.
22-218-19-01	489	Miocene	84
22-218-22-02	575	Miocene	72
22-218-25-02	700	Miocene	48
26-250A-10-06	100		5
27-259-09-05	100	U. Cretaceous	b.d.
27-263-06-05	150	U. Cretaceous	120
29-280A-10-04	165	Eocene	5
29-280A-15-02	280	Eocene	5
29-280A-19-00	450	Eocene	b.d.
29-281-11-00	90	Miocene	b.d.
29-281-14-00	130	Oligocene	110
31-299-22-01	190	Pleis-Pio	250
31-299-36-01	500	Late Miocene	80
31-302-10-00	180	Late Miocene	100

[a]Approximate figures as obtained from Site Summaries in the *Initial Reports of the Deep Sea Drilling Project.*

[b]The chlorin content was calculated using the 668-nm peak and $E = 3 \times 10^4$ l/mole per cm. Porphyrin was calculated using the Soret peak and $E = 3.3 \times 10^5$ l/mole per cm.

1. Cores frozen at -20°C
2. Ball mill extracted with 90% acetone-10% methanol
3. Sephadex LH 20 chromatography of extracts
4. Treatment of column fractions with diazomethane

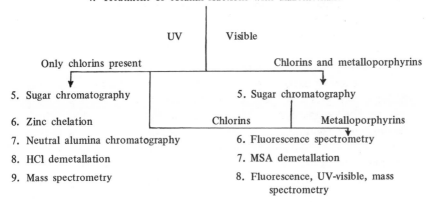

Figure 4.2 Experimental flow sheet.

fractions, as determined by UV-visible spectroscopy, eluted from the Sephadex column were treated with diazomethane to convert any free acid functions to esters. The diazomethane was prepared in a test tube in an ice bath by stirring together 3 ml 50% potassium hydroxide, 7 ml ether, and about 0.1 g N-methyl-N-nitro-N-nitrosoguanidine. The tetrapyrrole pigment in 5 ml ether was then slowly added to the diazomethane (at around 5°C) and stirred for 5 min, after which the solution was neutralized with 20 ml of 5% hydrochloric acid and the layers separated. Next the etheral layer was washed three times in 20 ml cold water and dried over anhydrous sodium sulfate.

Rechromatography of the esterified pigments over sugar (80-100 mesh) with cyclohexane-benzene fractionated the tetrapyrroles into the respective chlorin and porphyrin classes.

During analyses of the early cores received from DSDP, extreme difficulty was encountered in separating the free base tetrapyrrole (chlorins in the case of Leg 4-Leg 15) from a "hydrocarbon impurity" which tended to block out the visible spectra at high energy. However, the laboratory-prepared metallochlorin derivatives could be chromatographed to yield a pigment fraction of suitable purity for spectrometry.[8]

Following this procedure, the chlorin pigments were treated with zinc acetate in acetone and chromatographed over neutral alumina with benzene and benzene/methanol (0.5-2.0%) to give pigment fractions having increased

purity.   After demetallation with 18% hydrochloric acid, the pigments
were subjected to electronic and mass spectral analysis.

Chlorins were not detected in the Lower Miocene and Cretaceous samples.   The porphyrin tetrapyrroles extracted from these cores were isolated
by repetitive chromatography over sugar and neutral alumina.   Where in-
sufficient sample for mass spectrometric analysis was isolated, the pigment
was analyzed by fluorescence spectrometry.

The core extract was treated with 5 ml of methanesulfonic acid (MSA)
for each gram of extract concentrate in a ball mill containing 16 glass
marbles, rolling at low speed for 4 hr at 105 ± 5°C.   After 4 hr, a mixture of 5 g ice and 5 ml water per gram of extract was added with
stirring.   The solution (ca. 50°C) was filtered and extracted with dichloromethane in a liquid-liquid extractor.   After conversion of the dication to
the free base by treatment with sodium acetate (100 ml, 10% w/v), the
porphyrin was transferred to ether.   The porphyrin was then partitioned
between the organic solvent and 6 N hydrochloric acid, and the acid
solution analyzed for dication fluorescence.   Sample size of standard compounds was approximately 10 ng, and the instrument parameters were set
at a moderate sensitivity level.

Mass Spectrometry

The mass spectra were obtained using an AE1 MS-9 spectrometer fitted
with a direct probe sample injection system.   The pigment was transferred
to the probe with a drop of benzene or THF.   After removal of solvent
with the rough pump, the sample was quickly injected and the spectrum
recorded.   Spectra were obtained at 70 ev and 12 ev at about 225 to
325°C.

Electronic Absorption Spectrometry

The UV-visible spectra were obtained using a Beckman ACTA C III
ultraviolet-visible linear wavelength spectrophotometer and a Beckman DK-2
spectrophotometer.   An Aminco-Bowman 8202 spectrophotofluorometer
was employed for fluorescence measurements.

RESULTS

Spectra of Fossil Pigments

The visible spectra of pigments extracted from core samples from
Leg 15 Site 147 indicated all are of the chlorin type (see Table 4.2).
Each of the sample extracts was chromatographed on Sephadex LH-20 to

Table 4.2 Electronic and Mass Spectral Data

DSDP Core	Mass Spec (amu)	Electronic Spec (nm)	Tetrapyrrole Type
14-144A-02-06		408,660	Chlorin
		392	Nickel porphyrin
14-144B-03-06		408,660	Chlorin
		392	Nickel porphyrin
15-147B-01-03	890,872,858,844,830	408,667	Chlorin
	626,624	400,410,667,672	Chlorin
15-147B-02-03	872,858,830	408,667	Chlorin
15-147B-05-00	872,858,830	408,667	Chlorin
15-147B-05-06	874,872,858,830	408,667	Chlorin
	626,624	408,410,667,672	Chlorin
15-147B-07-04	874,872,858,830	406,663	Chlorin
15-147B-09-00	872,858,830	406,663	Chlorin
15-147C-02-03	874,872,858,830	406,663	Chlorin
15-147C-05-02		405,658	Chlorin
20-198A-04-03		b.d.	NFC[a]
22-217-02-04		410,663	Chlorin
22-217-10-04		b.d.	PFC[b]
22-217-27-02		b.d.	PFC[b]
22-217-32-02		b.d.	PFC[b]
22-217-32-02		392,550	Nickel porphyrin
22-218-05-02		408,663	Chlorin
22-218-11-02		b.d.	PFC[b]
22-218-19-02		408,660	Chlorin
		394,525,555	Nickel porphyrin
22-218-22-02		408,660	Chlorin
		395,525,555	Nickel porphyrin
22-218-25-02		405,658	Chlorin
		396,520,558	Nickel porphyrin
26-250A-10-06		392	Nickel porphyrin
27-259-09-05		b.d.	
27-263-06-05		389,513,551	Nickel porphyrin
29-280A-10-04		394	Nickel porphyrin
29-280A-15-02		393	Nickel porphyrin
29-280A-19-00		b.d.	NFC[a]
29-281-11-00		b.d.	NFC[a]
29-281-14-00		408,666	Chlorin
		395	Nickel porphyrin
31-299-22-01		408,660	Chlorin
		400,495,528,563,620	Prophyrin
		395,553	Nickel porphyrin
31-299-36-01		400,485,528,563,620	Porphyrin
		395,552	Nickel porphyrin
31-302-10-00		408,663	Chlorin
		400,495,528,563,620	Porphyrin
		394,520,555	Nickel porphyrin

[a]Negative check by fluorescence.
[b]Positive check by fluorescence.

yield four distinct bands: (1) dark green, (2) red, (3) green and (4) yellow, each having the chlorin-type electronic spectrum. The absence of major bands in the 600- to 640-nm region suggests that metallochlorins are not present. The appearance of Soret bands at or above 400 nm and the absence of characteristic peaks from 500 to 600 nm exclude pigments of the porphyrin type,[4] at least as major components of the mixture. However, the intense chlorin bands may well obscure the porphyrin components present in very small amounts. Also, bacteriochlorins do not seem to be present; otherwise the red band would be at much lower energy (about 750 nm).

After rechromatography of the red fraction of sample 15-147B-1-3 over sugar and then over neutral alumina, the collected fraction gave a 70-ev mass spectrum with peaks at m/e 872, 858, 844 and 830. Under low-voltage conditions (10 ev) the major peaks were at m/e 872 and 858. The electronic spectrum, major bands at 668 and 410 nm, indicates the vinyl group and carbonyl functions are intact, suggesting a dihydropheophytin structure.

The partial 70-ev mass spectrum of the green pigment fraction of sample 15-147B-1-2 is shown in Figure 4.3. Parent ion peaks were observed at m/e 626 and a small peak at m/e 624. The visible spectrum of this sample exhibits bands at 672, 668, 410 and 400 nm. Of the possible structures, chlorin p_6 trimethyl ester correlates best with the observed spectral data, since reduction of the 2-vinyl group is known to produce a blue shift of the red band from 672 to 668 nm. That the m/e 624 and 626 peaks in the mass spectra correspond to tetrapyrrole parent ions is supported by their increase to a series at m/e 685, 687 and 689 upon laboratory conversion to the copper derivatives. The presence of a copper chelate is also indicated by a shift of the 668-nm visible band to 635 nm.

The separation scheme followed for the isolation of pigment fractions did not exclude all nonpigment hydrocarbons, as indicated by a hydrocarbon envelope at 517 to 603 m/e appearing in all mass spectra (see Figure 4.3). This "hydrocarbon impurity" also tended to block out the visible spectrum at high energy, necessitating the subtraction of these effects in the electronic and mass spectra in order to interpret the obtained data. In subsequent analysis of DSDP cores, a procedure was developed which afforded clean separation of the pigments.

The mass spectral data were obtained for the pigments after diazomethane treatment. Therefore, any free acid functions in the tetrapyrroles were converted to methyl esters which increased volatility.

The pertinent geological data and absorption spectra of the tetrapyrrole pigments extracted from sediments taken on Leg 22 are also given in Tables 4.1 and 4.2. Total pigment was separated into red and green

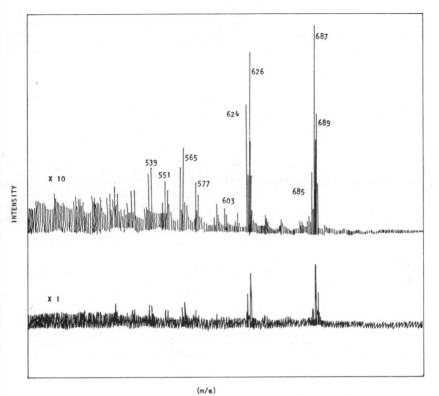

(m/e)

Figure 4.3 Partial mass spectrum of DSDP Leg 15, Site 147 pigments.

components by chromatography and, in line with conventional terminology, the green pigments have been called chlorins and the red ones, porphyrins.

The yield of pigment was very low from Site 217 cores, undetectable spectrophotometrically in three cores, and less than 10 ppb in the other two Site 217 samples investigated. The yield of tetrapyrrole was higher in Site 218, varying inversely proportional to core depth.

Samples of the green pigment from Sites 217 and 218 were found as free base chlorins. The absence of metallochlorin as in previous deep sea sediment studies[1,7] suggests that loss of magnesium from chlorophyll is a facile process, occurring at or before the pigment reaches the sediment-water interface.

The chlorins isolated from 22-218-5-2, an Upper Pleistocene sample, gave electronic spectra very similar to chlorins of Pleistocene age from Site 147 (Leg 15) with identical placement of the major absorption bands.[7] Combined electronic and mass spectral data indicated that pheophorbide (molecular weight 872, 858 and 830) and chlorin p_6 (molecular weight 890 and 624) type tetrapyrroles are present in Recent sediments.

The pheophorbide-like electronic spectra are consistent only with intact chromophores, again suggesting a dihydrophytol pheophorbide structure for the pigments of molecular weight 872 and 858. Pheophytin a and b are likely precursors.

The group of green pigments yielding chlorin p_6-type spectra (molecular weight 890 and 624) are proposed to arise by isocyclic ring opening of the dihydrophytol pheophorbide and pheophorbide from the reduction and elimination reactions of pheophytin. The observation of these compounds in Recent cores suggests that etioporphyrin series formation begins in early chlorophyll diagenesis in deep sea sediments.

The position of the electronic absorption red band of Site 218 chlorins shifts from 663 nm in the Pleistocene sample to 660 nm in Upper Miocene to 658 nm in Lower Miocene. This suggests that more complete reduction of ring-conjugating groups, such as the 9-keto or 2-vinyl, occurs with increased core depth and age.

Site 218 (Leg 22) Miocene cores yielded both green and red pigments. The major peaks at 395, 525 and 555 nm in the electronic spectra of the red fractions indicate the porphyrins are chelated with metal. Vanadyl petroporphyrins isolated from Athabasca tar sands gave a Soret band at 404 nm; nickel petroporphyrins from gilsonite gave a Soret at 387 nm. Baker et al.[4,5] have shown that these pigments consist of only chelated homologous series of alkyl- and cycloalkylporphyrins.

The observed data from Site 218 samples indicate that the chelating metal is nickel. The displacement of the Soret band from 387 nm for nickel alkylpetroporphyrins from gilsonite to 395 nm for Site 218 Miocene samples may be due to ring-conjugated substituents, such as the carboxylate group. In the Upper Cretaceous Site 217 pigment electronic spectrum, the Soret is observed at 392 nm, a blue *shift* toward that expected of a nickel alkylpetroporphyrin.

Each of the sediment samples investigated in the series of cores obtained from Legs 14, 20, 26, 27 and 29 was low in tetrapyrrole pigment concentration (see Tables 4.1 and 4.2). Samples 27-263-6-5 and 29-218-14-0 yielded approximately 115 ppb pigment, whereas most of the other cores analyzed gave tetrapyrrole yields 5 ppb or lower.

The youngest sample of this suite of cores (29-281-14-0, Oligocene age) contained both major classes of tetrapyrrole: chlorin and porphyrin. The electronic spectrum of the core extract consisted of a band in the red region at 666 nm and two major bands in the Soret region (*ca.* 400 nm). A band at 408 nm is taken to indicate the presence of chlorin pigment while the band at 395 nm may be ascribed to metalloporphyrin.

Leg 29 Eocene samples gave electronic spectra characteristic of metalloporphyrin with the Soret appearing at 394 nm.

The absorption spectra obtained for samples from this series of cores followed the pattern observed in previous DSDP core analyses.[1,7,8]   The Oligocene to Late Eocene samples exhibited both chlorin and metalloporphyrin peaks in the visible spectra.   The older sediments (Upper Cretaceous) contained no detectable free base chlorin, only tetrapyrrole belonging to the metalloporphyrin class.

Samples containing less than 5 ppb pigment (see Table 4.1) were subjected to treatment with methanesulfonic acid (MSA) to remove chelating metals and enhance the pigment fluorescence, thus lowering the detection limit.   Each of the extracts was heated with MSA, diluted, cooled and neutralized with saturated sodium acetate under an ether phase.   The porphyrins were then extracted from the ether with a small volume of 6 N hydrochloric acid.

The fluorescence spectra of all porphyrin extracts and reference compounds (mesoporphyrin 1X dimethyl ester and etioporphyrin I) were obtained through excitation at the 390- and 405-nm range and scanning the 550- to 700-nm emission range for the two principal dication fluorescence peaks.

As indicated in Table 4.2, the presence of tetrapyrrole was detected in all but three of the sediment core samples investigated by fluorometry.

The analytical results of core 27-263-6-5 are of special interest due to the relatively high yield of pigment which permitted more extensive analyses to be carried out.   One-half of the core sample made available for investigation in our laboratory (360 g) was exhaustively extracted with 90% acetone-10% methanol.   The visible spectrum indicated the presence of approximately 43 μg of nickel porphyrin.

After treatment with diazomethane, the pigment was chromatographed on neutral alumina to yield two fractions.   The fraction eluted with benzene-methanol (about 35 μg) had UV-visible peaks at 551, 513 and 389 nm.   The more polar fraction eluted with tetrahydrofuran absorbed at 548, 510 and 391 nm.   Since the first fraction contained most of the pigment, it was selected for treatment with MSA.   The tetrapyrrole was demetallated so as to determine its porphyrin structure using characteristic absorption patterns in the 500- to 600-nm region of the electronic spectrum.

The pigment was stirred with highly purified MSA for 4 hr at 100°C in a test tube.   Without cooling, the mixture was diluted with ice and water and filtered with suction.   The MSA solution was then extracted with dichloromethane in a liquid-liquid extractor for 16 hr.

Due to a high-level background, interpretation of the UV-visible spectrum of the dichloromethane extract was not possible.   This solution was then shaken twice with 10 ml 4% hydrochloric acid in order to extract the porphyrin from the interfering neutral organic compounds.   However, very little free base tetrapyrrole was recovered from the neutralized extract. Shaking with 18% hydrochloric acid gave an extract which, after neutralization under ether with sodium bicarbonate, gave a visible spectrum with peaks at 648.5, 595, 530, 505 and 395 nm, very unlike a typical free base porphyrin (see Figure 4.4a).

In addition to the tetrapyrrole recovered from hydrochloric acid extraction of the dichloromethane solution, a small amount of pigment was obtained when the MSA solution was further treated with sodium carbonate under ether. The electronic spectrum of the concentrated ether indicated the presence of peaks at 620 and 565 nm in addition to the 648, 595, 530, 505 and 395 nm peaks. Extraction of the ether solution with 3% hydrochloric acid removed the compound(s) absorbing at 620 and 565 nm, leaving a solution with an identical spectrum to that of the dichloromethane extracts.

The pigment recovered from the 6 N hydrochloric acid extract had a chlorin-type absorption spectrum with peak 1 at 649 nm. The overall spectrum appearance was very similar to that obtained for a compound isolated from a petroleum porphyrin aggregate by Fisher and Dunning.[11] They reported the pigment to be a noncarboxylated material which resisted conversion from chlorin to porphyrin upon treatment with quinone by Eisher's method,[27,28] thereby indicating that the material was probably not a chlorin.

Suspecting that the compound(s) was an artifact, the dioxy derivative of mesoporphyrin 1X dimethyl ester (dioxy-meso 1X DME) and the dioxy derivative of deoxophylloerythrin methyl ester (dioxy-DPE ME) were prepared by the method of Fischer.[29] As shown in Figure 4.4a, the spectra of dioxy-meso 1X, dioxy-DPE, and the extracted compound(s) are very similar.

In addition, the three compounds exhibited similar spectral changes, noticeably the appearance of a peak at 615-630 nm, when treated with diazomethane to convert free acid functions to methyl esters (see Figure 4.4b).

The unknown sediment pigment and the two dioxy-model compounds were treated with nickel acetate in acetic acid-acetone to produce the metal chelates. Again the results revealed a close similarity in spectra, especially between the unknown compound and the dioxy derivative of deoxophylloerythrin (see Figure 4.4c).

Lembert[30,31] reported that the hydroxyporphyrins or oxyporphyrins have a chlorin-like spectrum, and this material may very well be an oxidation product, formed upon prolonged standing by benzene solutions of porphyrins. Therefore, considering the possibility of oxidation, caution must be exercised in the selection of acidic reagents for demetallation or

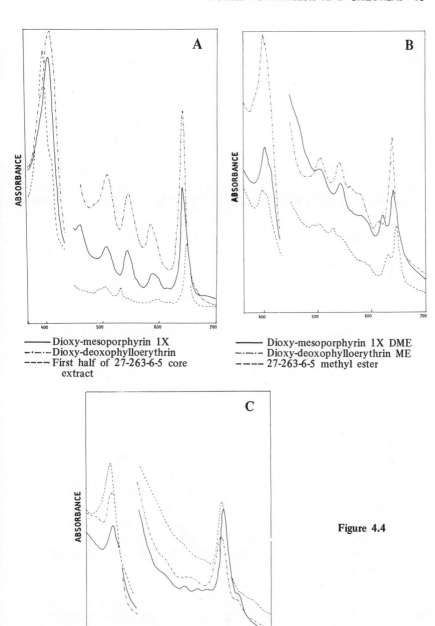

A

—— Dioxy-mesoporphyrin 1X
—·—·—Dioxy-deoxophylloerythrin
————First half of 27-263-6-5 core
      extract

B

—— Dioxy-mesoporphyrin 1X DME
—·—·— Dioxy-deoxophylloerythrin ME
———— 27-263-6-5 methyl ester

C

—————— Nickel dioxy-mesoporphyrin 1X DME
—·—·—·— Nickel dioxy-deoxophylloerythrin ME
———— Nickel 27-263-6-5 methyl ester (first half)

Figure 4.4

for extraction of porphyrins from bitumens. The most widely used reagent for the demetallation of porphyrin aggregates in petroleum and petroleum residues is hydrobromic acid-acetic acid (HBr-AcOH). However, any free bromine in the HBr-AcOH reagent will produce artifacts, as mentioned by many workers. For example, Chang et al.[32] reported that treatment of deutero- or protoporphyrin in HBr-AcOH gave rise to a compound of the chlorin class when bromine was present.

Baker et al.[5] reported satisfactory yields in the extraction of asphaltenes with methanesulfonic acid, but destruction of the porphyrin when attempts were made to demetallate either purified vanadyl petroporphyrin fractions or vanadyl porphyrins with specially purified MSA.[6] However, the addition of hydrazine sulfate to the reaction mixture produced essentially quantitative yields of free base porphyrin in each case.

Comparative results with the model dioxy compounds indicated that during MSA treatment of the first half of the extracts of core 27-263-6-5 oxidation to the dioxy derivative occurred. The experimental procedure was then repeated using the second half of the core with precautions taken to prevent oxidation.

The remaining core (360 g of 27-263-6-5) was extracted and the isolated pigment treated with technical methanesulfonic acid and about 50 mg of phenylhydrazine. After heating for 2.5 hr at 105°C the solution was worked up as before. The recovered free base tetrapyrrole had an acid number below 4, unlike the dioxy-porphyrin which had an acid number of around 15. Comparison of the electronic spectrum with standards suggested a mixture of DPEP-type and Etio-type free base porphyrins (see Figure 4.5a).

Reaction of the porphyrin mixture with nickel acetate gave a product having an electronic spectrum very similar to that of the nickel chelates of various synthetic porphyrins and identical with that of the pigment isolated from the original core acetone extracts (see Figure 4.5b). Again, the spectrum of the nickel derivative of the sediment porphyrin indicated a mixture of DPEP-type and Etio-type tetrapyrroles. A comparison of the relative intensities of the α- and β-bands points to such a mixture (see Table 4.3).

Inspection of the total collected data leads to the speculative identification of the 27-263-6-5 Upper Cretaceous core pigments as predominantly nickel deoxophylloerythrin and closely related nickel phylloporphyrins with a lesser amount of other carboxylated nickel porphyrins. The existence of such a mixture is consistent with the report by Treibs[2] of the identification of carboxylated compounds in petroleum porphyrin aggregates.

Observation of the intact carboxylic acid is significant, since the presence of such a functional group indicates a low-temperature history. Treibs[2]

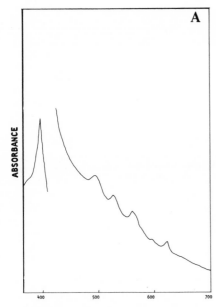

————Electronic spectrum of porphyrin extracted from second half of 27-263-6-5.

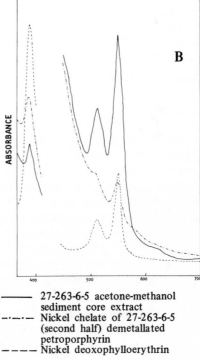

———— 27-263-6-5 acetone-methanol sediment core extract
—·—·— Nickel chelate of 27-263-6-5 (second half) demetallated petroporphyrin
———— Nickel deoxophylloerythrin

Figure 4.5

Table 4.3   Nickel Porphyrin Absorption Data

Compound	α-Band (nm)	β-Band (nm)	α/β Ratio
Nickel DPEP[a]	555	522	2.09
Nickel Deoxophylloerythrin	551	512	2.55
Nickel Etioporphyrin[a]	555	526	3.01
Nickel Mesoporphyrin DME	553	517	3.36
27-263-6-5[b]	551	513	2.78
27-263-6-5[c]	551	513	2.65

[a]Isolated from a sample of gilsonite.
[b]Acetone-methanol extract of sediment core.
[c]Nickel chelate of free base porphyrin isolated from MSA extracts.

has shown that porphyrin carboxylic acids lose carbon dioxide when subjected to temperatures above 200°C in the laboratory for extended periods of time.

The tentative identification of the chlorin-like compound as a dioxy by-product of the separation scheme is of substantial importance. These findings could possibly explain the origin of many of the secondary chlorins reported by various workers.[2,16,33]

DISCUSSION

Our investigations of the cores available for organic geochemical analysis from the Deep Sea Drilling Project have revealed much information concerning early diagenetic reactions of chlorophyll. The sediment core samples studied have fairly well defined geochemical histories and provide significant and consistent information on the intermediate products and mechanisms of the degradation pathways of chlorophyll diagenesis.

We have found no magnesium-chelated chlorins in the deep sea sediments examined. This suggests that the initial step in the diagenesis of the chlorophylls in open ocean environments is the loss of magnesium. Demetallation is complete at or near the sediment-water interface and may even occur during the fall through the water column. We do not, however, discount the possibility of chlorophylls surviving to burial in special environments, such as shallow water basins, fiords, and deltas where the sedimentation rate and plankton production are very high, and especially where reducing conditions exist.[21]

Nissenbaum *et al.*[34] detected chlorophyll *a* in sediments from the Dead Sea, which they attributed to slow degradation of organic matter and to high concentrations of magnesium in the water.

Great care must be exercised during extraction and workup to prevent formation of a metallochlorin which, on the basis of its electronic spectrum, could be mistaken for chlorophyll. This chelate was demonstrated by mass spectrometry to be the copper chelate, and in subsequent improved extraction procedures to be an artifact. With this in mind, we urge caution in uncritical acceptance of reports of the presence of unaltered chlorophyll in deep sea sediments.

Data collected from the analyses of sediment cores of Pleistocene age suggest the presence of chlorin p_6 compounds. Pigments isolated from the Cariaco Trench cores gave electronic spectra consistent with chlorin p_6 and molecular weights of 626 and 624 as indicated by mass spectrometry. We propose that these pigments are derivatives of chlorophyll and are formed by ring opening of the allomerization product of pheophytin.

Allomerization consists of oxidation of pheophytin and pheophorbides to give products having C-10 alkoxyl or hydroxyl substituents. Laboratory experiments indicate that the allomerization product of methyl pheophorbide undergoes chemical degradation and leads through a purpurin 18 intermediate to chlorin p_6[35] (see Figure 4.1b). Our observation of chlorin p_6 in deep sea sediment suggests that allomerization is a significant diagenetic reaction, and a mechanism by which the isocyclic ring can be opened leading to Etio and Phyllo porphyrins. These series of tetrapyrroles (those which do not contain an isocyclic ring) account for a considerable percentage of the porphyrins occurring in petroleum and other bituminous materials.[4,5,36,37]

It is possible, from a strictly structural chemical point of view, that members of the heme family of porphyrins are the precursors of the fossil Etio porphyrins (see Figure 4.6). However, as Corwin[38] has pointed out, this possibility is not very likely since the relative amount of material from plant sources is so overwhelming, perhaps even 100,000 to 1. Therefore, the major chlorophylls and the minor pigments of the heme type in the plant sources most likely contribute more to the overall store of the petroporphyrins than the major pigments of animal origin.

The observation of only chlorin-type pigments in Recent deep ocean sediments (see Table 4.2) is supportive evidence for this postulate. The absence, or existence in concentrations below visible absorption detection limits (3-5 ppb), of hemin derivatives in the recently deposited sediments suggests a minimal contribution to fossil porphyrins by the hemin precursors.

In addition to the proposed pathway for the formation of Etio porphyrins by the allomerization of chlorophyll derivatives, sufficiently rigorous

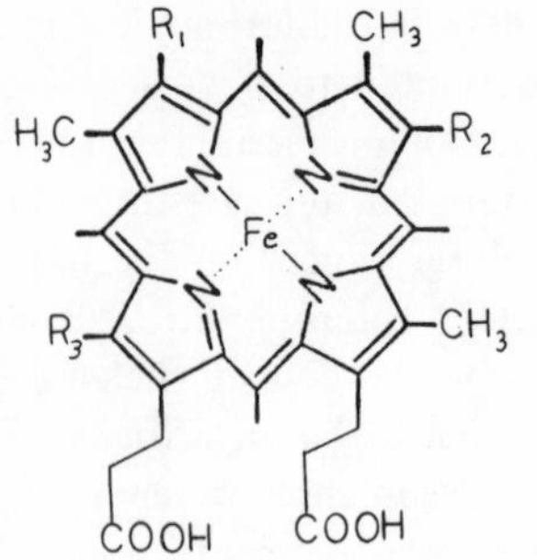

Name	R_1	R_2	R_3
Protoheme IX	$CH=CH_2$	$CH=CH_2$	CH_3
Chlorocruoroheme	CHO	$CH=CH_2$	CH_3
Heme c	V_c	V_c	CH_3
Heme a	V_f	$CH=CH_2$	CHO
Deuteroheme	H	H	CH_3

$V_c = CH(CH_3)SCH_2CH(NH_2)COOH$
$V_f = CHOHCH_2(CH_2CH=C-CH_2)_3H$
$\qquad\qquad\qquad CH_3$

Figure 4.6   Structures of heme compounds.

thermal conditions will open the isocyclic ring of compounds of the DPEP-type. For example, the porphyrins from retorted shale oil contained a much higher ratio of Etio to DPEP series than porphyrins recovered from the unretorted shale.[6] Baker[4,5] reported that the ratio of Etio to DPEP porphyrin was considerably higher in gas oil fractions which had been subjected to high distillation temperatures than in the original crude. Thermal opening of the isocyclic ring of DPEP is also consistent with the observation of high ratios of Etio to DPEP porphyrins in crudes of high thermal stress, such as Baxterville and Burgan.[4,5]

We have noted chromatographic separation of spectrally identical (electronic and mass) pigments extracted from organic-rich deep ocean sediments. It seems likely that we are dealing with mixtures of C-10 diastereomers. However, since equilibration of the diastereomers occurs so readily, it is impossible to ascertain whether these are artifacts of our purification scheme or are native to the sediments. Katz and co-workers found that equilibration between chlorophyll a and chlorophyll a' (C-10 epimers) is established quickly in THF at $10°C$.[39]

Given a sediment age of up to 80,000 years, it seems likely that epi-merization occurs prior to laboratory workup. Since epimerization produces only structural changes, not compositional change in the molecule, one may ask if it is even geochemically significant. It is possible, as with allomerization, that products produced by coupling reactions leading to higher-molecular-weight species arise through epimerization. Such coupling reactions, if they can be demonstrated, would have geochemical significance.

In previous papers[1,7,8] we have pointed out our observation of intact ester linkages in pigments isolated from sediments of Pleistocene and Late Pliocene age. These observations seemed to contradict the generally held assumption that hydrolytic deesterification occurs very early in the diagenesis of chlorophyll. Recent investigations have reinforced our earlier work and reopened the question of when deesterification occurs and by what mechanism.

We continue to observe intact ester linkages for Pleistocene cores from the Cariaco Trench.[7] These results taken in combination with the isolation of a methyl pheophorbide from an Eocene brown coal by Dilcher et al.[40] suggest that hydrolysis of pheophorbide esters is lethargic.

Equally suggestive results have been reported regarding the alcohol, i.e., phytol. In a study of the isoprenoid alcohols in recent and older sediments, Sever and Parker[41] failed to find significant amounts of phytol, but did find its reduction product, dihydrophytol. More recently, Simoneit[42] has reported that the phytadienes (predominately neophytadiene) represent approximately 1% of the solvent-soluble organic portion of a deep ocean sediment.

There seems little doubt that the phytol in the chlorophyll pigments is the primary source of these phytadienes. A positive correlation plot of phytadiene vs. chlorin pigment concentration lends modest support to this suggestion (see Figure 4.7).

Proposing a straightforward mechanism for the formation of the phytadiene, though, leads to some puzzling contradictions. One is forced to propose an essentially concurrent hydrolysis (water + pheophytin → pheophorbide + phytol) and dehydration (phytol → phytadiene + water) to account for the observed products.

The above leads us to suggest that elimination rather than hydrolysis is the common deesterification route (pheophytin → pheophorbide + phytadiene). Detailed consideration of the elimination mechanism predicts neophytadiene as the only product. The basis for this prediction can be found by examination of models showing the presumed geometry of the transition state. In Figure 4.8 is shown schematically the required bond rearrangements. Note that in the cis-isomer (natural phytol) only the

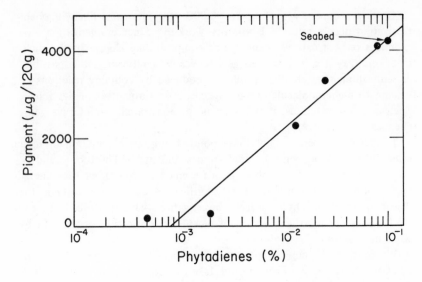

Figure 4.7 Total tetrapyrrole pigment yield versus phytadiene yield.

protons of the 3-methyl group can approach the carbonyl oxygen; hence neophytadiene is the predicted product. Since neophytadiene is thermodynamically the least stable of the phytadienes, analyses showing it to be the major component of the sediment phytadienes[42] lend powerful support to the suggested mechanism. Other mechanisms such as clay-catalyzed phytol dehydration are predicted to lead to mixtures in which the predominant isomer is the most thermodynamically stable.

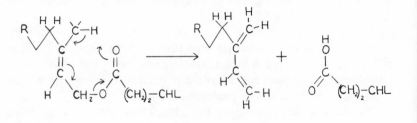

| Pheophytin | Neophytadiene | Pheophorbide |

Figure 4.8 Proposed mechanism for deesterification of pheophytin (neophytadiene—thermodynamically least stable of the phytadienes).

The elimination mechanism proposed (see Figure 4.8) would obviously be inoperative in the absence of the 2,3 *double* bond. For the reduced analogues, suitable transition states can be drawn, but they lack the driving force of a conjugated product and considerably more severe conditions would be required for the reaction to produce detectable amounts of products.

That the elimination reaction is inhibited by reduction of the phytol double bond could explain the apparent absence of pheophytin *a* (M.W. 870) and presence of dihydropheophytin (M.W. 872) in deep ocean sediments. That is to say, if reduction occurs, the ester linkage remains intact and one observes dihydrophytol pheophorbide *a* (M.W. 872). Conversely, if reduction is disfavored the ester linkage is ruptured by the elimination reaction with pheophorbide and phytadiene being formed.

Since methyl esters cannot undergo elimination (no olefin can be formed) the suggested mechanism also is consistent with the observed retention of the C-10 carbomethoxyl group in the pheophorbides (see Figure 4.1a).

If hydrolysis of the ester linkages were to be invoked, then one must reckon with the fact that the methyl ester linkages would be hydrolyzed at a greater rate than the phytyl ester. In turn, facile decarboxylation of the β-keto acid would produce the pyro series to the exclusion of pheophorbides, which is contrary to observations showing the pheophorbides to be dominate in the early sediments.

Assuming that the chlorophyll pigments have not undergone isocyclic ring opening shortly after burial, where the pigments are subjected to reducing environments, reduction of three functional groups is possible, considering chlorophyll *a* pigments which comprise the majority of the marine chlorophylls. Of the possible reduction sites in chlorophyll *a*, 2-vinyl, 9-keto, and the phytyl double bond, the phytyl olefin function undergoes reduction first.

Pigments isolated from several deep sea sources yielded spectral data consistent only with a dihydropheophytin *a* structure. A molecular weight of 873 (M.W. of pheophytin *a* is 870) proved by the mass spectrum and the 668-nm band in the electronic spectrum suggest only an intact chromophore (see Figure 4.9). More specifically, the total electronic spectrum is very similar to pheophytin *a* (see Figure 4.10). These data tend to rule out reduction of the vinyl group or formation of ring-conjugated carbonyl functions as likely structures.

That the compounds of molecular weight 830 and 858 are derived from 872 is supported by the observation that the percent of 872, relative to 830 and 858, decreases with increasing age of sample throughout the Pleistocene.

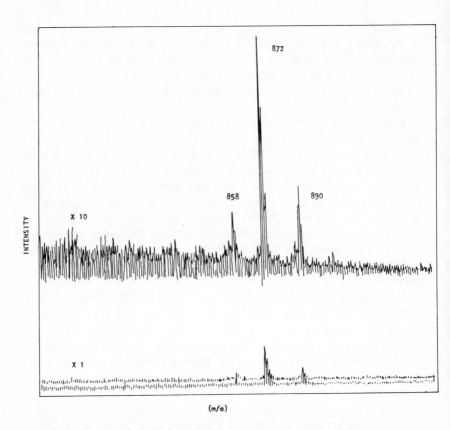

Figure 4.9 Partial mass spectrum of pigment fraction from DSDP Leg 15, Site 147 core.

Under more rigorous reducing conditions or with extended geochemical time, the 2-vinyl group can be converted to an ethyl substituent. Presumably reduction of the vinyl olefin is more difficult than that of the phytyl due to the additional stability afforded the vinyl group by conjugation with the aromatic (tetrapyrrole) ring. Spectra of the meso compounds prepared by catalytic reduction are very similar to the parent compounds, but the long wavelength bands are displaced to the blue by 5 to 10 nm.

There are few reported examples of mesochlorins isolated from sediment sources. However, chlorin p_6 and mesochlorin p_6 were found in the green pigment fraction of a sediment sample from the Cariaco Trench.[7]

Blumer[17] reported the presence of mesochlorins in Triassic oil shale, but suggested these ancient chlorins are not direct descendants of

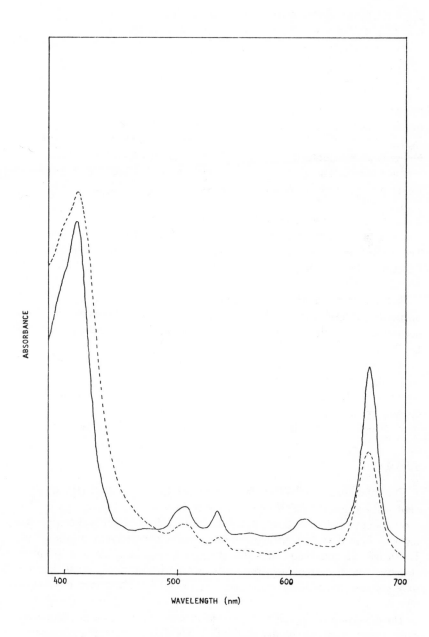

Figure 4.10 ———— Electronic absorption spectrum of pheophytin *a*.
– – – – Pigment fraction from DSDP Leg 15, Site 147 core.

chlorophyll, but rather secondary synthetic chlorins formed by hydrogenation of porphyrins under extreme reducing conditions.

At the third reduction site, the 9-keto group is converted to a methylene group. This reduction is a different type than the earlier ones. The first two reductions result in hydrogen enrichment of the molecule. In contrast, the third reduction results in a loss of oxygen, as well as a gain in hydrogen. To date no geopigments of a structure corresponding to a simple reduction of the 9-keto group have been reported and are not expected to be found in significant amounts. Presumably oxygen loss proceeds at a significant rate only under conditions sufficiently severe to produce elimination of nonconjugated olefins. Conditions this severe would also produce other reactions such as aromatization and possibly decarboxylation.

In the deep ocean sediment samples the spectrum of structures from barely altered chlorophyll to metal petroporphyrins have been found. The earliest appearance of metal porphyrins was reported in a Pleistocene-Pliocene sediment, and the oldest free base chlorin found was in a Middle Eocene sediment.[8]

Between these limits generally a mixture predominantly composed of unchelated chlorin and metalloporphyrin is observed. However, the occurrence of small amounts of free base porphyrins was detected in Leg 31 Miocene samples (see Table 4.2).

It appears that the chlorins undergo substantial reduction of ring-conjugating functional groups before the chelating metal is introduced. Apparently, only a small fraction of the chlorins undergo isocyclic ring opening prior to metal insertion, for the free base porphyrins and initially formed metalloporphyrins consist predominantly of DPEP type.

SUMMARY

The investigation of 32 selected Deep Sea Drilling Project (DSDP) core samples gave results which point to the mechanisms and pathways of the early geochemical transitions of chlorophyll in deep ocean sediments. Indeed, these new data strongly support the original scheme outlined by Treibs for the conversion of chlorophyll to thermally stable petroporphyrins. However, the results of the core analysis suggest that the process is more *complex* and demand that additional reactions be grafted onto the Treibs scheme.

The first step in the chemical degradation of chlorophyll in marine environments appears to be *demetallation*. The loss of magnesium occurs very early in the diagenetic pathway, possibly before burial.

Shortly after burial, the chlorophyll-containing debris is subjected to reducing environments. Stepwise *reduction* of tetrapyrrole ring substituents correlates with increasing geologic age and depth of burial.

Interpretation of core data in conjunction with previously reported results leads us to propose that *deesterification* proceeds in diagenesis through an *elimination* reaction. The elimination is inhibited by reduction of the phytyl double bond, in which case deesterification proceeds at a geochemically significant lower rate.

It is apparent that the elimination-reduction competition is a crucial *branch point* in chlorophyll diagenesis. The relative solubility of the pigment in aqueous and nonaqueous phases, as well as absorption parameters, are determined at this point. The free carboxyl function of pheophorbide *a* confers a modicum of water solubility, surface activity and an absorption site. Subsequent reaction of pheophorbide probably occurs either at water-oil interfaces or on solid surfaces, thus giving a quite different slate of products than the hydrophobic esters. With the ester linkage intact, oxidation, reduction and elimination reactions typical of nonaqueous environments are predicted.

Allomerization of chlorophyll derivatives in the recent sediments is proposed as a significant geochemical reaction. Opening of the isocyclic ring by this mechanism leads to chlorin p_6 tetrapyrroles. The chlorin p_6 compounds are possible precursors of Etio- and Phyllo-type fossil porphyrins and, therefore, allomerization should be considered as a route to alkyl porphyrin formation.

In addition to this mechanism the conversion of DPEP to Etio- and Phyllo-type porphyrins probably occurs under severe geothermal conditions by carbon-carbon bond rupture in the strained carbocyclic ring.

The geologic age span of chlorin to porphyrin *conversion* appears to be broad, ranging from Pleistocene to Middle Eocene. The observation of only nickel as the chelating metal in porphyrins isolated from a wide range of DSDP sediment samples suggests a selective mechanism for chelation of tetrapyrroles in deep ocean sediment environments.

Under more severe conditions, greater depth of burial and extended time, the porphyrins undergo diagenetic *transalkylation* reactions which lead to homologous series of alkylporphyrins including members of carbon skeleton number beyond those which are accountable by the mechanisms proposed in the original Treibs scheme.

ACKNOWLEDGMENTS

This research was supported by the Oceanography section of the National Science Foundation under NSF Grants Ga-31712 and GA-37962.

Appreciation is expressed to Dr. Heinz G. Boettger of the Jet Propulsion Laboratory of the California Institute of Technology and Mr. Jim Boal of the Mellon Institute for assistance in obtaining the mass spectra.

Our sincere thanks are extended also to Dr. Phillip Jobe of the School of Pharmacy of Northeast Louisiana University for help in obtaining the fluorescence spectra.

REFERENCES

1. Baker, E. W., C. Dereppe and J. R. Boal. *Preprints, Div. of Petrol. Chem. ACS* 15, A7 (1970).
2. Treibs, A. *Ann. Chem.* 509, 103 (1934); *Ann. Chem.* 510, 42 (1934); *Angew. Chem.* 49, 682 (1936).
3. Hodgson, G. W., B. L. Baker and E. Peake. In *Fundamental Aspects of Petroleum Geochemistry*, B. Nagy and U. Colombo, Eds. (New York: Elsevier, 1967), p. 242.
4. Baker, E. W. *J. Am. Chem. Soc.* 88, 2311 (1966).
5. Baker, E. W., T. F. Yen, J. P. Dickie, R. E. Rhodes and L. F. Clark. *J. Am. Chem. Soc.* 89, 363 (1967).
6. Baker, E. W. In *Organic Geochemistry*, G. Eglinton and M. T. J. Murphy, Eds. (New York: Springer-Verlag, 1969), p. 479.
7. Baker, E. W. and G. D. Smith. *Initial Reports of the Deep Sea Drilling Project*, Vol. XX (Washington: U.S. Government Printing Office, 1973), p. 943.
8. Smith, G. D. and E. W. Baker. *Initial Reports of the Deep Sea Drilling Project*, Vol. XXII (Washington: U.S. Government Printing Office, 1974), p. 667.
9. Vernon, L. P. and G. R. Seely. *The Chlorophylls*. (New York: Academic Press, 1966), and references therein.
10. Dunning, H. N. and J. W. Moore. *Bull. Am. Assoc. Petrol. Geologists* 41, 2403 (1957).
11. Fisher, L. R. and H. N. Dunning. *U.S. Bur. Mines, Rept. Invest.* 5844, 19 (1961).
12. Dunning, H. N. and J. K. Carlton. *Anal. Chem.* 28, 1362 (1956).
13. Dunning, H. N., J. W. Moore, H. Bieber and R. B. Williams. *J. Chem. Eng. Data* 5, 546 (1960).
14. Sugihara, J. M. and R. M. Bean. *J. Chem. Eng. Data* 7, 269 (1962).
15. Sugihara, J. M. and L. R. McGee. *J. Org. Chem.* 22, 795 (1957).
16. Blumer, M. *Anal. Chem.* 33, 1288 (1961).
17. Fox, D. L. and L. J. Anderson. *Proc. Natl. Acad. Sci. U.S.* 27, 333 (1941).
18. Fox, D. L., D. M. Updegraff and D. G. Novelli. *Arch. Biochem.* 5, 1 (1944).
19. Shabarova, N. T. *Biokhimiya* 19, 156 (1954).
20. Vallentyne, J. R. *Can. J. Bot.* 33, 304 (1955).
21. Orr, W. L., K. O. Emery and J. R. Grady. *Bull. Am. Assoc. Petrol. Geologists* 42, 925 (1958).
22. Hodgson, G. W., B. Hitchon, R. M. Elofson, B. L. Baker and E. Peake. *Geochim. Cosmochim. Acta* 19, 272 (1960).

23. Hodgson, G. W. and B. L. Baker. *Bull. Am. Assoc. Petrol. Geologists* **41**, 2413 (1957).
24. Hodgson, G. W. and B. L. Baker. *Nature* **202**, 125 (1964).
25. Hodgson, G. W. and E. Peake. *Nature* **191**, 766 (1961).
26. Hodgson, G. W., B. L. Baker and E. Peake. *Res. Council Alberta, Inform. Ser.* **45**, 75 (1963).
27. Eisner, U. *J. Chem. Soc.* **3749** (1955).
28. Eisner, U. and R. Linstead. *J. Chem. Soc.* 1655 (1956).
29. Fischer, H. and H. Orth. *Die Chemie Des Pyrrols: Part II* (Leipzig: Akad, Verlag, 1937), p. 273.
30. Lembert, R. and B. Cortis-Jones. *Biochem. J.* **32**, 171 (1938).
31. Lembert, R. and J. Legge. *Hematin Compounds and Bile Pigments* (New York: Interscience Publishers, Inc., 1949), p. 91.
32. Chang, Y., P. Clezy and D. Morell. *Australian J. Chem.* **20**, 959 (1967).
33. Blumer, M. *Helv. Chim. Acta* **33**, 1627 (1950).
34. Nissenbaum, A., M. J. Baedecker and I. R. Kaplan. *Geochim. Cosmochim. Acta* **36**, 709 (1972).
35. Conant, J. B., J. F. Hyde, W. W. Mayer and E. M. Dietz. *J. Am. Chem. Soc.* **53**, 359 (1931).
36. Dean, R. A. and E. V. Whitehead. Paper V-9, 6th World Petroleum Congress, Frankfurt, Germany (1963).
37. Thomas, D. W. and M. Blumer. *Geochim. Cosmochim. Acta* **28**, 1147 (1964).
38. Corwin, A. H. Paper V-10, 5th World Petroleum Congress, New York (1959).
39. Katz, J. J., G. D. Norman, W. A. Svec and H. H. Strain. *J. Am. Chem. Soc.* **90**, 6841 (1968).
40. Dilcher, D. L., R. J. Pavlick and J. Mitchell. *Science* **168**, 1447 (1970).
41. Sever, J. and P. Parker. *Science* **164**, 1052 (1969).
42. Somoneit, B. R. *Initial Reports of the Deep Sea Drilling Project*, Vol. XXI (Washington: U.S. Government Printing Office, 1973).

5

PRELIMINARY INVESTIGATION ON THE PRECURSORS OF THE ORGANIC COMPONENTS IN SEDIMENTS–MELANOIDIN FORMATIONS

D. K. Young, S. R. Sprang and T. F. Yen

   Departments of Environmental Engineering,
      Chemical Engineering and Medicine (Biochemistry)
   University of Southern California
   Los Angeles, California   90007

INTRODUCTION

The reaction of a carbohydrate with an amine group ($H_2 NR$) at elevated temperature generally results in the formation of melanoidin,[1] a dark brown-colored material.   The significance of this type of reaction is implicit in the formation of melanoidin in a wide variety of natural processes.   In the food processing industries an understanding of the mechanism of melanoidin formation would improve the control of deterioration and flavor development in foods.   Organic sediments found in our ecosystem are rich in possible raw materials for the melanoidin reaction;   for example nucleic acids, proteins, polysaccharides and mucopolysaccharides.   Therefore, it may also play an important role in the formation of humic acids under natural conditions.   On the basis of the similarities of the physical and chemical properties of the melanoidins and humic acids Maillard[1,2] and Enders and Theis[3] suggested that melanoidins could be precursors of nitrogen-containing humic acids.

Different starting materials as well as experimental methods have been used in attempts to elucidate the melanoidin reaction mechanism.   For instance, Haugaard, et al.[4,5] have followed the kinetic behavior of the D-glucose-diamino acids system while Song et al.[6-9] have investigated the inhibitory effects of bisulfite on the D-glucose-glycine system. To fully establish the mechanism, some means of separating and character-izing the end products are necessary. Qualitative paper chromatography was commonly used by previous investigators,[8-11] and partial identifica-tion of the separated products by specific chemical tests was carried out. This chapter reports the initial quantitative separation of the melanoidin reaction products from the model system of glucosamine in a neutral medium.

Drozdova[11,12] has demonstrated the formation of melanoidin from chitin (poly-N-acetylglucosamine) and glucosamine hydrochloride isolated from living organisms. Essentially, similar experimental procedures are followed in the present study to obtain melanoidin products that are comparable to the previous experiments. In particular, we would like to study the relative distribution of the low-molecular-weight products (<4500 daltons) by prolonged heating (up to about 100 hours) of the glucosamine system.

The reason for the study of the melanoidin formation at neutral pH for prolonged times, of course, is to simulate the conditions in our ecosystem. The understanding of the precursors of naturally occurring organic components in sediments is henceforth increased.

METHODS

All melanoidin reactions were conducted by refluxing, at $100°C$, each 500 ml sample of 0.1 N glucosamine hydrochloride (Calbiochem A grade) for the desired amount of time (1.5, 24, 73 and 102 hours). The initial pH was adjusted to neutral or slightly alkaline by the addi-tion of NaOH and measured for each sample (Table 5.1). The brown-colored solution from each reaction was lyophilized and redissolved in phosphate buffer (phosphate ~ 0.01 M, NaCl ~ 0.15, pH ~ 7.20) to make a solution at least four-fold concentrated. This concentrated solution was layered ($\sim 3\%$ of the total gel column volume) onto a

Table 5.1 Conditions of Melanoidin Reaction in Glucosamine System

Sample No.	Reaction Time (hr)	Initial pH
I	1.5	7.00
II	24	7.00
III	73	7.96
IV	102	7.70

Sephadex G-15 column (2.5 x 40 cm) previously equilibrated by the same phosphate buffer used before. The elution rate was about 0.6 ml/minute, and three ml fractions were collected. Before layering the concentrated reaction sample, the void volume of the gel column was determined by blue dextran 200.

The elution pattern was followed by absorbance measurement at 277 mn with a Beckman DB instrument. The UV spectrum of each of the eluted peaks was scanned with the same instrument (320-220 nm scan range, recording percent transmittance).

A resinous material displaying considerable activity at 277 nm was isolated as follows. The solution from Sample IV (refluxed for 102 hours) was stirred with 5.0 g of Norite activated charcoal and filtered through #50 Whitman Paper on vacuum-assisted Buchner funnel (5 g celite filtering aid was added to prevent Norite from contaminating the filtrate). The same procedure was repeated twice within the filtrate until the final filtrate was colorless.

The Norite from each treatment was combined and bathed in 100 ml N′N′-dimethylacetamide (DMAc), the mixture was warmed at 65-70°C for 12 hours. The entire mixture was transferred to a Mettler ultrasonic cleaner and sonicated for 15 minutes in a distilled water bath. The solvent (DMAc) was evaporated in a rotary evaporator at 75°C under low pressure (~ 0.01 torr). The resin thus obtained was redissolved in 10 ml DMAc. Ultraviolet spectra of the redissolved resin and the final filtrate were taken.

RESULTS AND DISCUSSION

The UV spectrum of the whole unseparated melanoidin product (Figure 5.1) from Sample IV (refluxed 102 hours) exhibits oval shape that is similar to those obtained by Drozdova[11] with short term (less than 5 hours) heating of the glucosamine system. In the present case, a strong absorption band is present around 280-285 nm (gram absorptivity ~2.1 x 10^4) and a minor shoulder is also detected at around 240 nm. It was suggested that the formation of intermediate products such as furfural and hydroxymethylfurfural[11,13,14] may be primarily responsible for the overall shape of the melanoidin spectrum. However, comparisons of the present melanoidin spectrum obtained by long-term heating of the glucosamine system with spectrum of either furfural or hydroxymethylfurfural[15] do not seem to support this idea. Indeed, as will be discussed later, compounds that are apparently stable with respect to prolonged heating may be contributing to the total melanoidin spectrum observed.

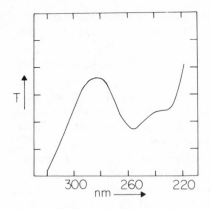

Figure 5.1   UV spectrum of total unseparated melanoidin reaction product of Sample IV (refluxed 102 hours).

Melanoidin products from each of the variously heated samples were separated on G-15 gel column and showed strikingly similar elution patterns.   The pattern of resolved peaks is well exemplified by Sample III, which was heated for 24 hours (Figure 2.5).   Essentially, four major components were resolved; they are designated as α, β, γ, and δ in accordance with the order of elution from the gel column.   Each of the components exhibited a unique UV spectrum (Figure 5.3).

The α component is the only one that exhibited a dark brown color;   presumably most of the melanoidin material is contained in this component.   The UV spectrum (Figure 5.3a) of this component is a very uncharacteristic hyperbolic absorbance increase from 320-220 nm. The dark brown (melanoidin) material apparently does not absorb in this wavelength region, since the Norite extraction removed practically all of the melanoidin material, leaving behind a colorless filtrate that showed an UV spectrum resembling the dark brown α-component.   The material extracted by Norite, on the other hand, has a strong absorption band around 280 nm that is comparable to the original unseparated reaction product.

The rest of the components (β, γ and δ) are all colorless, and this is where most of the UV absorption lies.   The β-component absorbs very strongly at around 275 nm and is probably the component that contributes significantly to the strong absorption band around 280-285 nm of the total melanoidin spectrum.   Again, a comparison of the spectrum with pure furfural and hydroxymethylfurfural[15] does not confirm the β-component as being either of these two compounds.   For instance $\epsilon_{max}/\epsilon_{min}$ is at least 10 for furfural or hydroxymethylfurfural,[15] and this is obviously not the case for the β-component, which has

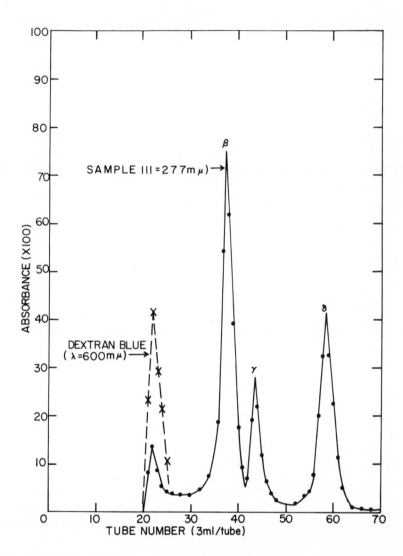

Figure 5.2 Chromatographic elution pattern of Sample III (refluxed 24 hours). The solid line represents Sample III taken at $\lambda = 277$ nm. The solution was concentrated 16 times (by lyophilization) and about 0.75 ml was layered on a 2.5 x 43 cm GC-15 gel column with a total pressure head of 53 cm. The buffer used was phosphate (0.01 M) and saline (0.15 N) at pH 7.2. The dashed line is dextran blue taken at $\lambda = 600$ nm eluted separately to determine the approximate void volume (V_o). About 0.4 ml of saturated dextran blue, dissolved in the same buffer as above, was layered on the column before the Sample III run.

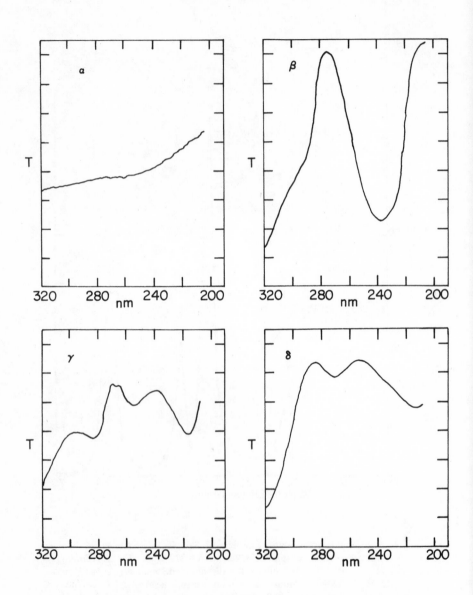

Figure 5.3   UV spectra of the isolated components (α, β, γ, and δ) of Sample III. In order of their elution they are α (tube number 22), β (tube number 38), γ (tube number 44), and δ (tube number 59).

$\epsilon_{275}/\epsilon_{240} \cong 6.97$. In addition to this, a smaller absorbance maxima at around 230 nm for furfural and hydroxymethylfurfural cannot be seen in the β-component spectrum.

Components γ and δ both absorb at more than one wavelength region showing moderately strong absorption bands. The minor shoulder observed at 240 nm for the total melanoidin spectrum (Figure 5.1) is likely to be the compilation of the absorption bands around this wavelength region of these two components.

The contribution of each resolved component was estimated by integrating the area occupied by the peaks, and the values for the samples heated for 1.5, 24, and 73 hours are shown in Table 5.2. It should be

Table 5.2   Relative Percent Distribution of Isolated Components at Various Reaction Times

Sample Number	α	β	γ	δ
I	7.58	48.50	17.63	26.29
II	9.71	52.04	16.60	21.65
III	9.08	59.80	18.45	12.68

pointed out that since we have followed the elution pattern at 277 nm only, this procedure for estimating the contribution of each component would tend to exaggerate the component that has a higher extinction coefficient around 277 nm while underestimating those components not showing significant absorption around this wavelength. Therefore, the values shown in Table 5.2 are not absolute percentage contributions, but are only meaningful values when comparing the relative distribution of each component with respect to the heating time of the reaction. The values in Table 5.2 indicate that the relative distribution of the four resolved components was established fairly rapidly, in no more than 1.5 hours (Sample I), and that the same distribution tends to persist on prolonged heating (up to 73 hours) with very gradual changes in the distribution. Components α and β both increased while the δ-component decreased with no significant change in the γ-component.

It was observed by Drozdova[11] that the color intensity of the melanoidin reaction increases rapidly during the first 5 hours then levels off to a slow steady state increase thereafter. The steady state increase would account for the slow changes or no changes in the component distribution after 5 hours of heating (i.e., at 24 hours and 73 hours) if the isolated components are transitory intermediates of the melanoidin reaction. However, we do not observe significant alteration of the distribution between 1.5 and 24 hours heating of the glucosamine system, and yet the brown-colored material (melanoidins) was shown to be rapidly accumulating after 1.5 hours of heating. This would argue against the assumption that these isolated components are transitory intermediates of melanoidin formation. In addition, the amount of starting materials as well as the conditions (except the reaction time for heating) are the same for Sample I (1.5 hours of heating) and Sample II (24 hours of heating) while more materials can be isolated from Sample II (judging from the absolute total area under all the isolated peaks). Therefore, it is more likely that these isolated components are relatively stable end-products of the melanoidin reaction.

The molecular weight of each of the isolated components could be estimated by its elution volume (V_e). For instance, Goodson et al.[16] have related V_e to molecular weight for G-15 gel. By defining the distribution coefficient (K_D)

$$K_D = \frac{V_e - V_o}{V_i}$$

where V_o is the void volume estimated from dextran blue elution and V_i is the accessible volume inside the gel matrix estimable from deuterium oxide elution, then

$$K_D = 1.537 = 0.421 \quad [\log_{10} MW]$$

This equation is valid only for molecules that do not have interaction with the dextran gel, for otherwise the elution volume (V_e) would be larger, resulting in underestimation of the MW. Values of K_D for each of the isolated components of Sample III are listed in Table 5.3.

Table 5.3 Values of K_D for Each of the Isolated Components of Sample III

Component	K_D	MW[a]
α	0	$\geqslant 4500$
β	0.54	$\geqslant 229.6$
γ	0.75	$\geqslant 72.4$
δ	1.28	—

[a]Calculated from $K_D = 1.537 - 0.421 \log_{10} MW$ (see Reference 16 and text for explanation).

There is obviously gel interaction in the case of the δ-component, since in the absence of such interactions K_D 1 would mean that molecules smaller than 19 daltons do not participate in the chromatographic process;[16] therefore, a value of $K_D > 1$ would indicate a delay in elution of δ-component due to interaction with the gel matrix. Gel interaction cannot be excluded for β and γ components. However, by assuming insignificant gel interaction and using the value K_D, which is less than unity in both cases, it is possible to set a lower limit of the MW of these two components. In Table 5.3 the lower MW limits for β and γ are 229.6 and 72.4 respectively. This would put β-component somewhere between glucose (a monosaccharide) and maltose (a disaccharide). If this is true it would also mean that β-component cannot be furfural or hydroxymethylfurfural. The γ-component is approximately the molecular weight of a triose. The α-component is obviously a high-molecular component (> 4500 daltons), since this component was eluted together with the dark-brown (melanoidin) material, as we have mentioned previously. Therefore, the present results indicate the possible existence of a high-molecular-weight non-melanoidin component.

The use of G-15 gel column to separate the product of melanoidin reaction has, so far, shown the existence of apparently stable small-molecular components as well as a large non-melanoidin component from prolonged heating of glucosamine. Further characterization of each of the components would be necessary to elucidate the mechanism of the melanoidin reaction.

ACKNOWLEDGMENTS

Partial financial aid through Gulf Oil Corporation, Chevron Oil Field Company, A.C.S. Petroleum Research Fund 6272-AC2 and technical discussion with Dr. Sol Silverman of Chevron are appreciated.

REFERENCES

1. Maillard, L. C. *Compt. Rend.* **154**, 66 (1912).
2. Maillard, L. C. *Compt. Rend.* **156**, 1159 (1913).
3. Enders, C. and K. Theis. *Brennstoff-Chemie* **19**(15), 360 (1938).
4. Haugaard, G. and L. Tumerman. *Arch. Biochem. Biophys.* **65**, 86 (1956).
5. Haugaard, G., L. Tumerman and H. Silvestri. *J. Amer. Chem. Soc.* **73**, 4594 (1951).
6. Song, P., C. O. Chichester and F. H. Stadman. *J. Food Sci.* **31**, 906 (1966).
7. Song, P. and C. O. Chichester. *J. Food Sci.* **31**, 914 (1966).
8. Song, P. and C. O. Chichester. *J. Food Sci.* **32**, 98 (1967).

9. Song, P. and C. O. Chichester. *J. Food Sci.* **32**, 107 (1967).
10. Chichester, C. O., F. H. Stadtman and G. MacKinney. *J. Amer. Chem. Soc.* **74**, 3418 (1952).
11. Drozdova, T. V. *Biochem.* **22**, 449 (1957).
12. Manskaya, S. M. and T. V. Drozdova. *Dok. Akad. Nauk. S.S.S.R.* **96**, 569 (1954).
13. Kretovich, V. L. and R. R. Takareva. *Biokhimiya* **13**, 6 (1948).
14. Wolfrom, M. L., R. D. Shultz and L. F. Cavalieri. *J. Amer. Chem. Soc.* **71**, 3518 (1949).
15. MacKinney, G. and O. Temmer. *J. Amer. Chem. Soc.* **70**, 3586 (1948).
16. Goodson, J. M., V. Distefano and J. C. Smith. *J. Chromatog.* **54**, 43 (1971).

6

COMPOSITION OF POLLUTED BOTTOM SEDIMENTS
IN GREAT LAKES HARBORS

Bernice M. Katz

    Departments of Oceanography and
      Environmental Engineering
    Florida Institute of Technology
    Melbourne, Florida 32901

Raymond J. Krizek

    Department of Civil Engineering
    The Technological Institute
    Northwestern University
    Evanston, Illinois 60201

Paul L. Hummel

    Department of Civil Engineering
    University of Hawaii
    Honolulu, Hawaii 96844

INTRODUCTION

Although there is currently much concern over the dredging and disposal of polluted bottom sediments in the vicinity of deep-water ports, there is likewise much controversy[1-3] regarding the technical relevancy and adequacy of presently used criteria for classifying the pollution potential of these dredged materials. Among the reasons for this situation are: (a) the lack of known correlations between causes and effects, (b) the apparent insignificance of many parameters when applied to sediments, (c) the unavailability of standardized testing procedures applicable to dredgings, and (d) the expense associated with any sophisticated testing program. This chapter includes the test procedures that were used to analyze 75 samples of waters and solid dredgings taken from seven different Great Lakes harbors, a concise presentation of all results, and a general

111

evaluation of the concentration, variability and distribution of pollutants throughout various phases of the dredging and disposal cycle.

TEST PROGRAM

Thirty-three different chemical tests were conducted on 16 A samples (waters from rivers or overflow weirs of disposal sites), 5 B samples (turbid waters from dredge hoppers or near discharge pipes at disposal sites), 22 C samples (shallow bottom sediments or slurries from discharge pipes at disposal sites), 6 D samples (deeper bottom sediments or thick muds from disposal sites), and 26 E samples (firmer materials from land-fills at different disposal sites). These samples were taken from Chicago; Cleveland; Detroit; Green Bay; Milwaukee; Monroe, Michigan and Toledo, Ohio. A detailed description of the exact sources of all samples has been given by Krizek, Karadi and Hummel,[4] and the sampling procedures have been reported by Hummel and Krizek.[5]

TESTING METHODS

The testing procedures used in this work are generally those outlined in the twelfth edition of *Standard Methods for the Examination of Water and Wastewater* for the more fluid materials and *Chemistry Laboratory Manual for Bottom Sediments* for the more viscous and stiffer materials. However, due to the high concentration of interfering compounds in certain cases, modifications to the standard test had to be made. Brief descriptions of the testing methods employed herein are given below.

Total Solids, Total Volatile Solids, and Silica

To determine total solids, a sample of bottom sediment was added to a preweighed crucible, quickly weighed to avoid evaporation loss, and dried overnight at 100°C. After weighing, total volatile solids were obtained by igniting the sample at 600°C for one hour. The ignited sample was digested one half hour on a steam bath in a solution containing approximately 10:1 concentrated hydrochloric acid and concentrated nitric acid; it was then filtered into a preweighed gooch crucible, and the filtrate was used for the total phosphate and metal analyses. After the crucible was dried for one hour at 600°C, the weight of the remaining solids was taken to be a measure of the silica content in the original sample. One hundred cm^3 aliquots of the aqueous samples were used to determine total solids and total volatile solids as in the sediments, but silica was not determinable.

Total Phosphate

The filtrate from the total solids determination was brought to 100 cm^3 in a volumetric flask and used to determine the total phosphate. A suitable aliquot was treated with ammonium molybate reagent and stannous chloride and measured photometrically at a wave length of 690 mμ according to the procedure outlined in *Standard Methods*. This procedure was found to be reliable for waters containing sewage, sludge effluents and detergents, and complete recovery of the organophosphates could be attained.

Soluble Phosphate

A weighed sample of bottom sediment was completely mixed with boiling water and filtered with repeated washings; the soluble phosphates were then determined in a manner similar to that used for total phosphate. Suitable aliquots of water samples were analyzed without any pretreatment, as described above.

Metals

The filtrate from the total solids was also used to determine the concentrations of aluminum, cadmium, calcium, copper, total iron, lead, potassium and sodium by atomic absorption on a Jarrell-Ash atomic absorption spectrophotometer (Model 82-500). Standards were prepared by using the same ratios of cations in addition to phosphate and treating in the identical manner as the sample. In all cases except calcium, the standard curves of these cations were identical to the standard curves for each metal by itself in deionized water. In the calcium determination, lanthanum chloride and the sodium salt of ethylene diamine tetraacetic acid (EDTA) solutions were added to both the sample and the standard to prevent interferences from phosphate and aluminum. To circumvent the need for excessive dilutions to attain the minimum sensitivity of each cation, instrumental parameters were varied for the more concentrated cations. In the case of iron, where a less sensitive wavelength had to be used, tests were performed to ensure that none of the other cations interfered at this wavelength. The water samples were analyzed for cations without pretreatment.

Mercury and arsenic could not be determined by the above method, since these metals are lost during ignition in the total volatile solids determination; furthermore, the procedure most recommended (refluxing the sample with strong oxidizing agents) was found to be too time consuming and to require too much equipment. In the mercury analysis the sample was digested in 200 cm^3 Wheaton pressure bottles at 121°C for one half hour with concentrated nitric acid, and the entire contents were emptied into a reaction vessel and reduced to the ground state with stannous chloride. The mercury was then vaporized with nitrogen gas as a carrier and measured by atomic absorption spectrophotometry using the flameless technique. Similarly, in the arsenic analysis the sample was digested at 121°C for one half hour with a solution of potassium permanganate-sulfuric acid in a pressure bottle, filtered, and diluted to a specific volume; the concentration was determined by the usual flame aspirator procedure, but this was not as sensitive as desired. The water samples were analyzed for mercury by the flameless method without pretreatment at the beginning of the study, but later it was found that some waters contained organomercurial compounds in very low concentrations, and these were not identified in the untreated samples.

To determine the concentration of soluble iron, the sample of bottom sediment was mixed in boiling water, cooled to room temperature with nitrogen gas, and filtered by bubbling nitrogen gas through it into a 100 cm^3 volumetric flask, whose filtrate was then analyzed by atomic absorption, as for total iron. Despite all precautions to avoid air, the deeper bottom sediments still formed a precipitate by the oxidation of soluble ferrous ions to insoluble ferric ions. For the E samples the soluble iron was taken to be the sum of the ferrous iron and the iron that precipitated.

Ammonia Nitrogen and Total Organic Nitrogen

After the sample was neutralized, the ammonia was distilled into a boric acid solution, which was titrated with 0.02 N hydrochloric acid. If the ammonia concentration was too low for titration (as was the case with the water samples), the distillate was measured photometrically with Nesslar's reagent according to the procedure described in *Standard Methods*. For total nitrogen, a potassium sulfate-mercuric sulfate-sulfuric acid solution was added to the residue of the distillation and digested until all the carbon was removed; it was then diluted with ion exchange water, made alkaline with a sodium hydroxide-sodium thiosulfate solution, and distilled as in the ammonia determination. A special precaution was taken with the bottom sediments to ensure that sufficient acid was added so that a liquid remained and the flask did not overheat.

Nitrite Nitrogen

The nitrite nitrogen was determined according to the procedure outlined in *Standard Methods*, whereby nitrites react with sulfanilic acid and couple with naphthylamine hydrochloride to form a reddish-purple azo dye, which can be measured photometrically at a wavelength of 520 mμ.

Nitrate Nitrogen

Of the procedures described in *Standard Methods*, the zinc reduction method did not work, and the cadmium amalgam reduction, although sensitive, was quite lengthy. After considerable study, the technique of West and Lyles[6] was found to be the easiest and most effective. Chromotropic acid dissolved in concentrated sulfuric acid was added to 0.2-1.0 cm^3 of sample, after adding 2 cm^3 of antimony in concentrated sulfuric acid and 1 drop of urea sulfamic acid, and this was diluted to 10 cm^3 with concentrated sulfuric acid; not more than 3% water should be present. The yellow color was read photometrically at a wavelength of 420 $m\mu$. For both nitrite and nitrate nitrogen the water sample was analyzed directly. After a weighed sample of sediment was mixed with dilute acid, filtered into a volumetric flask, and brought to volume with repeated washings, a suitable aliquot was taken.

Sulfide

The sulfide content was determined according to the Titrimetric Method whereby a sample is treated with sulfuric acid and the resulting hydrogen sulfide is carried over with carbon dioxide bubbling through into a second and third flask containing a zinc acetate solution. The zinc sulfide was then reacted with an excess of standard 0.025 N iodine solution and back-titrated with standard 0.025 N sodium thiosulfate solution.

Oil and Grease

The oil and grease content of the more solid sediments was determined by a procedure whereby the sample is acidified with concentrated hydrochloric acid, dried to a powder with anhydrous magnesium sulfate and extracted with hexane in a Soxhlet extractor. Less solid sediments were likewise acidified with acid, but they were filtered on a Buchner funnel with muslin cloth, filter paper and filter aid preparation, as for sewage. Some of the samples had to be filtered as dry as possible, and anhydrous magnesium sulfate was added to the watery solids. The four-hour extraction with hexane was performed on most of the samples, but it was noted that this period was not long enough for some samples containing large concentrations of grease, so the extraction was continued until no more color was extracted. Oil and grease were recovered by evaporating the hexane, and the hydrocarbon content was determined by redissolving the residue in 10 cm^3 of hexane and passing this solution through an activated alumina column.

Phenols

To determine the phenolic content the sample was acidified, the phenols steam-distilled, and the distillate reacted with aminoantipyrine, potassium ferricyanide; after color development, this was extracted in a known amount of chloroform and measured photometrically at a wavelength of 460 mμ.

Chemical Oxygen Demand

A suitable aliquot, predetermined from the total volatile solids, was weighed into a 250 cm^3 flask, and mercuric sulfate was added; then 0.25 N potassium dichromate was added until any green color that formed was reversed. After adding concentrated sulfuric acid silver sulfate solution, the mixture was refluxed for two hours, cooled and titrated with standard ferrous ammonium sulfate; the difference between this and a blank was used to calculate the total chemical oxygen demand due to organic carbon and inorganic reducing agents. For the water samples 10 cm^3 of 0.25 N potassium dichromate and 20 cm^3 of sample were used.

Biological Oxygen Demand (BOD)

Sampling and the possibility of toxic material were the main problems encountered in determining the BOD of sediments. A small amount of sample and the sewage to be used for seeding were aerated overnight to acclimate the bacteria to any toxic substances that might be present. To ensure an even distribution of the mixture in the BOD bottles and sufficient sample for the proper dilution (due to the high organic content of the sample), a weighed sample was added to a four-liter aspirator bottle with a stirring bar, and continuous washings with the dilution water brought the sample to two or three liters, depending on the amount of dilution required. While mixing over a magnetic stirrer, three BOD bottles were filled, care being taken to avoid air bubbles; one was used to determine the initial dissolved oxygen, and two were placed in the 20°C incubator. Two dilutions of each sample were made. The 5-day BOD values for the two bottles with the same dilution usually differed by less than 1%, and the other two dilutions usually exhibited less than 5% difference. During incubation the bottles were mixed gently to ensure that there was no settling of the solids. Dilutions for the water samples were made by direct additions to the BOD bottles.

Hydrogen Ion Concentration, Oxidation-Reduction Potential, and Turbidity

The hydrogen ion concentration, pH, was measured by a Beckman expandomatic pH meter on the regular scale. Oxidation-reduction potential, Eh, was measured by the same instrument on the expandomatic scale with a platinum electrode. Turbidity readings were taken for the water samples on a Hach turbidimeter and read in JTU units. The meter was standardized with appropriate dilutions of a 150 JTU stock solution containing 300 mg/l Kaolin.

TEST RESULTS

The results of this test program are summarized in Table 6.1. Although it is considered that two significant figures can be used reliably for most of these test data, an attempt was made to establish columns of data

Table 6.1

Sample	Total Solids	Total Volatile Solids	Total Suspended Solids	Total Volatile Suspended Solids	Turbidity	Acidity	Hydrogen Ion Concentration (pH)	Oxidation-Reduction Potential (Eh)	Dissolved Oxygen (DO)	Biological Oxygen Demand (BOD)	Chemical Oxygen Demand (COD)	Organic Nitrogen (N)	Ammonia (NH_3) Nitrogen (N)	Nitrite (NO_2) Nitrogen (N)	Nitrate (NO_3) Nitrogen (N)	Total Phosphate (PO_4)	Soluble Phosphate (PO_4)
	%	%	%	%	JTU	mg/gm		mv	mg/gm	mg/gm	mg/gm	mg/gm	mg/gm	mg/gm	mg/gm	mg/gm	mg/gm
A6	0.0620	34.24	0.03	0	0.65		7.4	324	13.6	7.1			0.04	1.23	24.67	2.82	0.39
A8	0.0909	9.40	0.34	0	1.10		7.3	125	0.2	18.7			10.78	2.02	33.12	3.30	0.18
A9	0.0583	20.98	0.46	0	0.79		7.2	129	14.3	34.8			0.07	1.27	7.50	10.38	1.03
A10	0.0934	10.88	3.56	1.50	4.3		7.4	209	8.3		17.5	3.64	0.60	0	0.02	2.36	0.99
A11	0.0353	23.04	3.54	2.55	3.4		7.5	413	28.0		34.8	6.79	0.68	0.04	0.34	2.41	1.47
A12	0.0144	21.21	10.57	5.04	4.0		7.9	821	84.4		0	5.22	0	0.03	0	4.52	2.78
A13	0.0432	37.87	5.49	1.22	18.0		7.6	273	30.3		79.1	3.56		0.90	5.93	3.12	1.06
A14	0.0450	36.89	5.11	1.06	19.0		7.8	300	24.9			3.36		0.70	8.29	1.78	1.07
A15	0.0419	15.99	7.64	2.15	22.0		7.7	318			66.3					4.65	
A16	0.0402	35.32	4.73	1.62	15.0		7.6	341	31.8		53.0	3.92	0.70	1.22	10.02	3.11	1.19
A17	0.0431	39.44	3.36	0.81	13.0		7.6		29.7				0.97			4.52	1.16
A18	0.0438	34.25	5.94	1.37	20.0		7.8	311	28.1		45.5	3.71	0.09	0.84	6.92	3.42	0.01
A19	0.0433	32.22	14.09	3.28	27.0		9.3	307	25.8		0.9	2.84	1.94	1.32	0	3.46	2.24
A20	0.0420	28.96	12.63	3.81	34.0		10.2	311	30.5		1.2	3.03	0	1.32	0	2.14	3.22
A21	0.0402	30.74	9.83	4.23	24.0		8.9	324	28.1		0.7	3.90	0	1.49	0	3.73	2.86
A22	0.0390	28.68	6.42	2.18	19.0		9.0	336	27.7		0.3	3.42	0	1.42	0	3.34	2.95
B5	3.76	10.15					7.3				204.0	1.04	1.64			0.067	
B6	11.93	11.77					7.0				206.0	0.05	1.45			0.132	
B7	4.26	12.18					7.1				431.0	0.17	1.79			0.133	
B8	0.767	1.6	8.9	0.93	35.2		8.8				2.6	0.3	2.45			0.65	0.06
B9	0.070	17.7	51.9	3.90	250		8.9	198.8	15.7		0.6	1.8	3.54	0.53	0	1.43	1.00
C5	25.23	11.13									301.6	2.37	5.12			81.49	
C7	0.29	11.51									48.9	0.33	1.19			28.13	
C8	4.18	24.08							27.1		72.2	1.97	0.27			8.19	
C9	1.94	9.29									184.5	5.25	1.44			6.54	
C10	18.10	14.61									114.4	3.36	1.67			13.86	
C11	14.51	14.44									180.8	3.46	1.23			11.62	
C12	20.24	12.12									47.4	1.77	0.40			8.37	
C13	18.79	10.56									97.0	3.17	0.47			7.69	
C14	14.23	17.34									85.8	2.33				7.03	
C15	18.07	10.08									105.2	3.71	0.37			7.82	
C16	28.61	10.71									79.8	4.10	0.84			9.71	
C17	12.04	10.61									94.5	3.18	0.49			6.38	
C18	20.00	9.72									84.5	3.38	0.85			6.97	
C19	20.83	9.39									88.0	4.00	0.59			14.19	
C20	25.67	10.21									141.4	3.18	0.52			8.19	
C21	19.10	10.32									116.0	4.10	0.94			8.19	
C22	17.33	9.03														6.42	

Sample	Total Solids %	Total Volatile Solids %	Total Suspended Solids %	Total Volatile Suspended Solids %	Turbidity JTU	Acidity mg/gm	Hydrogen Ion Concentration (pH)	Oxidation-Reduction Potential (Eh) mv	Dissolved Oxygen (DO) mg/gm	Biological Oxygen Demand (BOD) mg/gm	Chemical Oxygen Demand (COD) mg/gm	Organic Nitrogen (N) mg/gm	Ammonia (NH₃) Nitrogen (N) mg/gm	Nitrite (NO₂) Nitrogen (N) mg/gm	Nitrate (NO₃) Nitrogen (N) mg/gm	Total Phosphate (PO₄) mg/gm	Soluble Phosphate (PO₄) mg/gm
C23	60.65	6.55															
C24	13.77	13.26															
C25	3.91	4.37															
C26	13.07	14.14															
C27	12.38	12.63															
D1	51.91	12.14	13.27														
D3	43.43	6.96	6.76														
D4	47.22	10.53	8.94														
D8	40.21	9.33										0.70	0.82			0.054	
D9	37.81	8.81										0.68	0.68			0.032	
D10	38.68	8.91										0.41	0.70			0.047	
E31	62.31	7.94									95.7	0	0	0.020	19.26	4.25	0.028
E33	56.48	6.16									100.9	244.1	360.2	0.009	33.40	3.70	0.038
E36	57.72	6.64									102.2	184.4				3.90	
E39	58.87	6.53									89.5	428.9	68.3	0.008	0	3.70	0.050
E40	65.82	9.89									117.8	120.6	0	0.008	0	6.00	0.053
E41	55.14	7.67									103.0	382.1	254.7	0.036	12.70	5.14	0.099
E43	58.17	6.92									69.6	49.4	145.7	0.009	0	3.84	0.040
E45	59.35	6.28									67.7	0	110.8	0.674	1.68	4.10	0.045
E48	65.22	8.78									122.8	235.9	0	0.019	44.54	5.86	0.036
E49	58.15	8.81									86.8	28.6	0	0.034	3.44	4.57	0.051
E52	53.49	7.81										23.2	99.8	0.834	5.56	5.20	0.159
E54	63.26	8.84										16.2	0	0.243	25.09	4.64	0.037
E55	64.90	7.50										81.3	0	0.238	4.62	3.42	0.022
E58	56.15	7.95										15.7	128.3			7.18	0.033
E60	69.19	8.41										108.1	0	0.286	2.51	7.48	0.028
E61	68.18	8.52										62.4	0	0.371	4.36	4.34	0.023
E63	72.31	8.56										27.2	75.5			5.11	
E65	64.47	9.14										56.5	0	0.464	0.37	7.97	0.024
E66	68.65	9.96										7.6	0	0.371	3.66	5.39	0.022
E68	64.16	8.07											116.0			5.41	
E69	55.91	10.56										12.3	383.1	0.797	5.31	5.70	0.064
E72	69.40	8.45										10.0	0	0.662	38.87	6.17	0.023
E73	72.09	8.38										46.8	0	0.620	27.52	4.61	0.021
E74	74.14	8.51										110.4	0	0.533	29.32	4.30	0.025
E75	70.02	8.28										71.5	95.4	0.564	28.22	2.98	0.031
E76	62.59	8.69										58.5	62.6	0.387	0.46	5.53	0.049

Sample	Grease mg/gm	Hydrocarbons mg/gm	Aluminum (Al) mg/gm	Arsenic (As) mg/gm	Calcium (Ca) mg/gm	Copper (Cu) mg/gm	Total Iron (Fe) mg/gm	Soluble Iron (Fe) mg/gm	Potassium (K) mg/gm	Silica (SiO₂) mg/gm	Sodium (Na) mg/gm	Sulfide (S) mg/gm	Cadmium (Cd) µg/gm	Cyanide (CN) µg/gm	Lead (Pb) µg/gm	Mercury (Hg) µg/gm	Phenolics µg/gm
A6			0	0	144	0	0.40		25.8		53.2		0		855	0.002	
A8			0	0	73	0	0		11.0		46.7		0		583	0.001	
A9			0	0	129	0	0.86		30.0		128.6		0		909	0.008	
A10			0		11	26.77	4.07	0.03	3.2		11.5		0		0	0.003	0
A11			0		29	104.73	3.20	0.07	7.6		56.6		0		0	0.013	0
A12			0		70	312.94	15.30	0	11.6		34.7		0		0	0.013	0
A13			0		238	3.70	8.80	0	12.5	0	41.7		0	0	0	0.001	3.7
A14			0		236	2.89	8.44	0	11.8	0	40.0		0	0	0	0.002	2.2
A15			0		263	2.15	8.12	0	14.6	0	52.5		0		0	0.001	
A16			0		274	3.23	17.41	0	14.2	0	54.7		0	0	0	0.001	4.0
A17			0		232	3.02	17.17	0	10.7	0	51.0		0		0	0.001	
A18			0		242	1.00	11.87	0	11.2	0	45.7		0	0	0	0.002	5.7
A19	11.55				370	0.51	12.01	1.16	11.1	0	76.2		0	0	0	0	0
A20	7.15				417	0.52	8.82	1.19	10.7	0	48.9		0	0	0	0.003	0
A21			0		448	0.82	3.73	1.24	11.7	0	57.3		0		0	0.002	62.2
A22			0		411	0.50	3.34	1.28	12.1	0	57.8		0		0	0.011	0

Sample	Grease	Hydrocarbons	Aluminum (Al)	Arsenic (As)	Calcium (Ca)	Copper (Cu)	Total Iron (Fe)	Soluble Iron (Fe)	Potassium (K)	Silica (SiO₂)	Sodium (Na)	Sulfide (S)	Cadmium (Cd)	Cyanide (CN)	Lead (Pb)	Mercury (Hg)	Phenolics
	mg/gm	mg/gm	mg/gm	mg/gm	mg/gm	mg/gm	mg/gm	mg/gm	mg/gm	mg/gm	mg/gm	mg/gm	µg/gm	µg/gm	µg/gm	µg/gm	µg/gm
B5	1.02	0.368	2.27		337.0	0.17	161.20		5.0	686	0.84	480	12.4	9.85	1340		
B6	0.21	0.070	16.09		380.0	0.11	31.51		4.6	825	0.66	594	16.0	8.13	106		
B7	5.71	2.300	17.90		625.0	0.11	194.00		4.5	644	1.01		14.9	18.55	161		4.16
B8			30.00		96.5	0.08	45.8	0.30	3.5	600	8.08		0		0		0
B9			23.00		186.5	0.60	14.3	7.17	11.5	340	28.00		0		0	2.65	0.71
C5			18.46		0.77	0.231	0.34		3.33	560.9	0.81		26.3		72		
C7	169.90		2.12		131.10	1.380	47.57		8.13	226.1	147.13		0	11.69	1413		
C8	11.40		12.23		43.59	0.781	44.24		1.57	351.1	10.66		0	5.97	428		
C9	25.00		14.22		62.18	0.485	23.91		4.03	237.7	13.34		77.4		191		
C10			14.61		19.72	0.155	29.44		3.77	634.4	0.53		9.7	11.82	144		
C11			19.34		23.88	1.061	29.97		4.28	634.3	0.62		10.2	4.06	150		
C12			18.62		15.44	0.850	28.51		4.99	661.4	0.60		10.7	3.26	161		
C13			18.88			0.639	26.45		4.52	677.3	0.63		8.6	1.12	101		
C14			17.30		21.47	0.058	25.98		4.39	682.9	6.76		9.6		145		
C15			17.85		32.79	0.053	26.12		5.11	685.5	0.65		6.1	1.11	81		
C16	4.26	1.36	16.09		21.09	0.049	25.45		4.61	689.0	0.47	63.4	5.3	3.84	84		
C17	0.83	0.24	18.86		29.09	0.068	27.85		4.95	673.6	0.62	240.3	7.3	3.41	129		
C18	4.42		17.82		22.31	0.064	30.57		4.83	693.5	0.58	19.6	3.6	0.95	99		
C19	1.17	0.44	16.89		26.13	0.052	28.08		5.06	687.1	0.50	54.8	7.0	2.30	87		
C20	11.90	6.81	16.77		24.13	0.042	31.83		5.05	521.4	0.53	14.6	9.7	2.81	105		
C21	1.36	0.84	15.86		23.09	0.061	29.45		5.44	675.9	0.50	92.8	9.1	5.76	102		
C22	11.48	5.71	16.28		28.99	0.051	30.94		5.34	686.7	0.55	28.1	9.0	1.15	73		
C23			8.82		0.78	0.039	22.94		1.29	803.1	3.66		16.6		89		
C24	5.30	2.86	18.23		36.98	0.356	15.84			624.5		294.2	0	5.52	178		
C25	11.10	6.39				0.677				844.7		191.6	0	10.98	161		
C26	9.16	4.94	19.43		35.50	0.384	15.84			612.4		316.0	0	7.50	181		
C27			24.15			0.443	19.47			597.0					181		
D1	36.40		4.01			0.106	0.52		0.38	502.9	0.73		9.0		61	0.853	2.7
D3	2.12		8.56	0.14		0.003	0.34		0.65	498.9	5.69		6.7		50	0.423	
D4	29.00		7.91	0.11		0.195	0.79		0.71	503.8	1.31		32.7		115	1.378	
D8	14.80	12.2	12.80			0.037	18.10		3.65	743.0	0.38	63.9	5.2	1.14	88		
D9	6.61	3.7	10.44			0.033	20.20		4.26	746.4	0.44	43.4	5.3	1.47	70		
D10	5.42	4.1	10.24			0.038	22.25		3.63	743.3	0.41	34.1	5.0	0.78	89		
E31	25.30	1.97	16.13	0	23.21	0.152	26.19	0.12	2.89	740.0	1.84	1.6	4.5		45	270	5.60
E33	0.46		13.79	0	25.69	0.194	23.64	0.26	2.62	697.5	1.28	57.5	2.5		41	370	0.20
E36	1.62	1.32	18.91	0	32.03	0.249	24.40		2.68	692.9	1.46	32.0	3.0		31	151	
E39	1.46		12.69	0	25.25	0.143	20.48	0.96	2.07	702.8	1.25	71.5	3.0		30	591	0.40
E40	2.42		20.36	0	20.38	0.270	31.25	0.37	3.06	676.4	1.60	1.8	5.7		44	213	0.40
E41	2.54	1.99	21.56	0	30.57	0.208	29.21	1.63	2.88	665.0	1.70	119.5	7.0		52	138	0.09
E43	3.52	1.02	15.16	0	22.82	0.076	21.07	0.26*	2.62	691.3	2.54	3.6	2.1		31		0
E45	1.77	0.47	15.20	0	21.54	0.108	22.15	0.66	2.86	688.1	1.45	5.9	3.1	0	38	123	0.47
E48	5.07	1.16	21.72	0	14.89	0.202	29.85	0.24	3.78	669.9	1.96	0.9	5.6		65	147	0.48
E49	2.56	0.49	18.35	0	19.37	0.198	25.10	0.36	3.11	683.0	2.04	2.0	3.8		50	127	0.48
E52	3.34	0.75	17.45	0	19.59	0.142	24.91	2.40	3.19	668.6	1.99		4.0		40	163	1.14
E54	2.74	0.63	20.05	0	13.88	0.166	30.06	0.21	2.98	686.4	1.84	1.7	4.7		59	158	0.76
E55	2.62	2.07	13.77	0	9.78	0.139	17.07	0.03	2.14	407.8	0.43	1.2	2.1		33	106	0.15
E58	2.90	2.72	28.44	0	32.42	0.343	40.93		5.49	661.8	3.80	7.1	7.1		90		0
E60	2.38	2.29	27.96	0	35.59	0.366	40.96	0.18	5.22	633.5	2.72	0.6	6.0		71	306	
E61	1.93	1.72	19.42	0	19.15	0.175	27.48	0.14	3.12	678.6	1.94	4.4	3.0		51	197	0.40
E63	4.20	2.94	19.56	0	20.30	0.157	25.20		3.58	727.0	2.37		2.8		43	180	
E65	3.35	1.91	17.99	0	20.58	0.261	26.48	0.09	2.73	674.8	1.68	1.2	5.5		55	160	0.40
E66	1.66	1.27	18.72	0	15.40	0.214	22.94	0.11	2.78	667.3	1.96	0.6	3.7		45	133	0
E68	1.08	10.93	18.49	0	23.96	0.252	27.39		2.96	666.2	1.96	3.1	3.7	0	51	344	0.40
E69	3.12	0.23	27.97	0	26.89	0.236	34.96	0.26	4.30	663.6	3.39	29.0	5.4		67	250	0.40
E72	2.18	1.51	20.96	0	19.01	0.231	27.60	0.03	2.66	643.4	1.85	0.9	4.7		45	134	0.50
E73	0.96	0.95	18.44	0	19.86	0.137	25.93	0.01	3.12	672.4	1.23	0.8	4.0		34	129	0.40
E74	1.43	1.04	15.00	0	25.19	0.162	23.37	0.02	2.49	677.4	1.14	1.4	3.4		38	239	0.40
E75	1.46		14.51	0	16.69	0.154	19.87	0.16	2.22	664.9	1.27		2.0		30	148	0.40
E76	1.58		17.40			0.212	27.06	0.29	3.09	660.0	1.70	101.6	3.5		48		0.20

according to a consistent decimal format, and this sometimes caused the number of significant figures to vary within that column. Insofar as possible, units were selected to maintain consistency. Total solids is expressed as a percentage of the weight of the solids to the total wet weight of the sample; all other weight ratios are expressed as the weight of the constituent to the weight of the solids present. Although this is admittedly unusual in the case of water samples, for which concentrations are normally expressed as the weight of the constituent per unit volume of sample (usually a liter), this format is employed to facilitate direct comparisons among the various other concentrations that can be expressed more appropriately on a dry weight basis. Furthermore, the EPA pollution criteria given in Table 6.2 for determining the acceptability of dredge spoil for open water disposal are given on a dry weight basis. The conversion to a wet weight basis is accomplished simply by multiplying the dry weight basis for a constituent by the ratio of the total solids to the total sample, taking proper account of the units of each.

Table 6.2   Guidelines for Limiting Concentrations of Various Pollutants in Bottom Sediments

Sediments in Fresh and Marine Waters	Concentration Percent (Dry Weight Basis)
Volatile solids	6.0
Chemical oxygen demand (COD)	5.0
Total Kjeldahl nitrogen	0.10
Oil and grease	0.15
Mercury	0.0001
Lead	0.005
Zinc	0.005

RESULTS AND DISCUSSION

Although details to support the subsequent interpretations can not be deduced from the data in Table 6.1 without knowing the specific sources of the samples,[4] this testing program was sufficiently broad in scope to give a general appreciation of the chemical composition of Great Lakes bottom sediments and the associated waters and to allow some interpretation of the interrelationships among the constituents found in the various phases of the dredging and disposal operation. For example, comparisons between the measured concentrations of the various pollutants and the

EPA standards given in Table 6.2 indicate that most of these dredgings would indeed be classified as polluted. The degree of pollution varies from harbor to harbor, from sample to sample within a harbor, and in the intensity of the various pollutants for a particular sample. The chemical composition of the water and bottom sediments was very consistent with that expected at particular locations (*e.g.*, where sewage was emitted, the dissolved oxygen was low and the biological oxygen demand and nitrogen content were high). However, the chemical oxygen demand depended not only on organic compounds, but also on other materials (such as the sulfide anions and the mercurous, cuprous and ferrous cations) that are found in high concentrations in these dredgings.

Several series of tests were conducted to investigate specific aspects of the dredging and disposal cycle; these include (a) the variation in water quality with distance from the outflow weir of a disposal area, (b) a comparison of constituent concentrations in the water from the river, dredge hopper and outflow weir, (c) composition of the sediments on the harbor bottom as a function of depth, and (d) the constituent concentrations during various phases of the dredging cycle. In general, the various constituents of water samples taken from different locations relative to the outflow weir of a disposal site indicated slight decreases in concentrations with distance from the weir for some constituents (such as total solids, total suspended solids, total iron, and aluminum) and a random distribution for most other parameters.

The percentage of solids in the water from the bins of a hopper dredge is much greater than that in the water from the outflow weir of the disposal area; the latter, in turn, is greater than its counterpart in the river water. The total volatile solids, total suspended solids, acidity, and concentrations of phosphates, calcium, copper, potassium and sodium were higher in the river water than in the water from the outflow weir, but the chemical oxygen demand, total iron and silica were higher in the water from the outflow weir than in the river water.

The deeper (1-3 ft) sediments (sampled by a flap-valve sampler)[2] exhibit about twice the amount of solids per unit wet weight of sample as the shallow (top 6 inches) sediments (sampled by an Ekman dredge).[2] The deeper sediments have a lower percentage of volatile solids and nitrogen and a much lower concentration of phosphates (on the order of one percent of the phosphates in the surface materials). Concentrations of oil and grease, as well as the metals tested, were about the same order of magnitude in both deep and shallow sediments.

The following observations concerning constituent concentrations during various phases of the dredging cycle may be advanced. The percentage of total solids per unit weight of wet sample are on the order of a few hundredths of one percent for the water samples, several percent for the water from the hopper of the dredge, 15-20% for the material from the discharge pipe and for the soft bottom sediments obtained by the Ekman dredge and flap valve sampler, and 50-70% for the fill materials. The percentage of total volatile solids per unit weight of total solids decreases from the water to the bottom sediments and decreases further in the fill materials. There was little variation in the chemical oxygen demand of the various materials, except for the water, which had a somewhat smaller value. Respective concentrations of organic and ammonia nitrogen were about the same in A and C materials, but they were significantly higher in the E or fill materials. Concentrations of total phosphates were about the same order of magnitude in all materials, but they were slightly lower in the water than in the bottom sediments and fill materials. The soluble metal ions, such as sodium and potassium, were found in greater concentrations in the water, whereas greater amounts of the heavy metals were found in the sediments and fill materials.

SUMMARY

Based on data (together with the associated testing procedures) from almost 2000 chemical analyses on 75 samples ranging from waters to solids from seven harbors in the vicinity of the Great Lakes, the pollution potential of dredged bottom sediments from deep-water ports and the variations in pollutant concentrations during various stages of the dredging and disposal operation were evaluated.

ACKNOWLEDGMENT

This work was supported in large part by the U.S. Environmental Protection Agency under Grants 15070-GCK and R-800948.

REFERENCES

1. Boyd, M. B., R. T. Saucier, J. W. Keeley, R. L. Montgomery, R. D. Brown, D. B. Mathis and C. J. Guice. "Disposal of Dredge Spoil: Problem Identification and Assessment and Research Program Development," Technical Report H-72-8 (Vicksburg, Mississippi: U.S. Army Engineer Waterways Experiment Station, 1972).
2. Keeley, J. W. and R. M. Engler. "Discussion of Regulatory Criteria for Ocean Disposal of Dredged Materials: Elutriate Test Rationale and Implementation Guidelines," Miscellaneous Paper D-74-14 (Vicksburg, Mississippi: U.S. Army Engineer Waterways Experiment Station, 1974).

3. Krizek, R. J. and G. M. Karadi. "Disposal of Polluted Dredgings from the Great Lakes Area," *Proc. First World Congress on Water Resources,* **4**, 482 (1973).
4. Krizek, R. J., G. M. Karadi and P. L. Hummel. "Engineering Characteristics of Polluted Dredgings," Technical Report No. 1, EPA 15070-GCK and R-800948 (Evanston, Illinois: Department of Civil Engineering, The Technological Institute, Northwestern University, 1973).
5. Hummel, P. L. and R. J. Krizek. "Sampling of Maintenance Dredgings," *J. Testing Eval.* **2**(3), 139 (1974).
6. West, P. W. and G. L. Lyles. "A New Method for the Determination of Nitrates," *Analyt. Chim. Acta* **23**, 227-232 (1960).

7

THE DISTRIBUTION OF HEAVY METALS IN
SEDIMENT FRACTIONS FROM MOBILE BAY, ALABAMA

J. M. Brannon, J. R. Rose, R. M. Engler and I. Smith

Department of the Army Corps of Engineers
U.S. Army Engineer Waterways Experiment Station
Environmental Effects Laboratory
Vicksburg, Mississippi 39180

INTRODUCTION

The distribution of trace metals in the various fractions of a sediment is of interest to sediment and water chemists for a number of reasons. A quantitative knowledge of the selective distribution of trace metals in sediments can be of value in assessing the potential impact of sediment resuspension upon water quality. The selective distribution of trace metals may also be helpful in evaluating the relative availability of these trace metals to influence biological communities and enter into reactions and transformations.

A large amount of information is available on the distribution of trace metals determined by various operationally defined selective extraction schemes,[1-4] but these procedures all subject the sediment to drying and grinding before chemical fractionation. The oxidation and physical alteration of the sediment during drying and grinding may cause phase or fraction differentiation of elements within the sediment.

The transformations of many elements under reducing conditions are important, but iron and manganese transformations in sediments can be particularly important due to the large amounts present and the scavenging effect of the hydrated oxides of iron and manganese on trace metal ions in solution.[5-8] Iron and manganese are solubilized by reducing conditions[9-12] and may release associated trace metals into solution.

The sediments studied to date for trace metal distribution involving chemical fractionation have consisted mainly of deep-sea sediments, with little attention given to estuarine systems. However, Sholkovitz[13] considered rapidly deposited sediments of near-shore basins an excellent environment to relate processes of early diagenesis to interstitial water chemistry. Estuarine sedimentation is much more rapid than in near-shore basins and should provide a good picture of interstitial water chemistry in a dynamic environment. By investigating other sediment fractions in addition to the interstitial water phase, it should be easier to understand the mobilization and migration processes that can occur in early diagenesis.

In this study, the distribution of iron, manganese, zinc and copper was studied in an operationally defined selective extraction scheme that precluded atmospheric oxidation and physical modification of the sediment during critical steps. The functionally defined sediment phases investigated included the interstitial water, adsorbed (ion exchangeable) on sediment material, reducible (solubility and migration controlled by oxidation-reduction reactions), bound in organic matter and residual (bound in interlayer positions of clay minerals or in a mineral crystalline lattice).

METHODS

The study area was Arlington Ship Channel in the heavily industrialized Mobile Bay and adjoining estuarine system of southern Alabama. Arlington Channel is located south of Mobile as shown in Figure 7.1. Five sediment cores, 60 cm in length and 7.5 cm in diameter, were taken at each of two sites. Site 1 was near the main ship channel, while site 2 was further west near the harbor turning basin.

Three cores taken at each site were immediately sealed in their polycarbonate core liners with polycarbonate caps, all of which had been acid washed. These cores were then stored upright in ice for transportation and subsequently stored at $4°C$. The two remaining cores were used for on-site measurements of pH, temperature and redox potential using a combination glass electrode, a YSI* temperature probe, and a bright platinum-calomel half cell, respectively. Platinum electrodes were inserted in each core to a depth of 6 cm, the core was sealed and the electrode allowed to equilibrate for 12 hours before redox potential (electrode potential) was measured.

---

*Citation of trade names does not constitute an official endorsement or approval of the use of such commercial products.

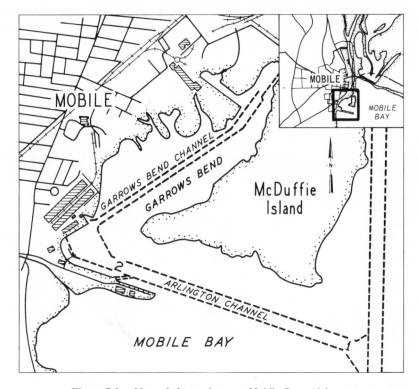

Figure 7.1   Map of the study area, Mobile Bay, Alabama.

Conductivity, salinity, temperature, depth, dissolved oxygen, and pH were measured in the water column in site 1 using an Interocean probe. Samples of the water immediately overlying the sediment were taken with a 5-liter polycarbonate Van Dorn water sampler. The water was then filtered on-site under about 100 psi N_2 gas in a polycarbonate filtration apparatus through a 0.45-μ pore size membrane filter. The first 250 ml of filtrate was discarded, and the remainder stored in a prerinsed polycarbonate bottle, acidified to pH 1 with HCl, and packed in ice until stored at 4°C.

The three sealed cores from each site were subjected to the sediment fractionation scheme (Figure 7.2) of Engler et al.[14]   Each 60-cm-long core was divided into four 15-cm sections before fractionation. The fractions investigated included the interstitial water fraction and the fraction extractable in 1 N ammonium acetate at the ambient sediment pH (exchangeable phase). The interstitial water and exchangeable phases were extracted under oxygen-free conditions. The next phase, designated

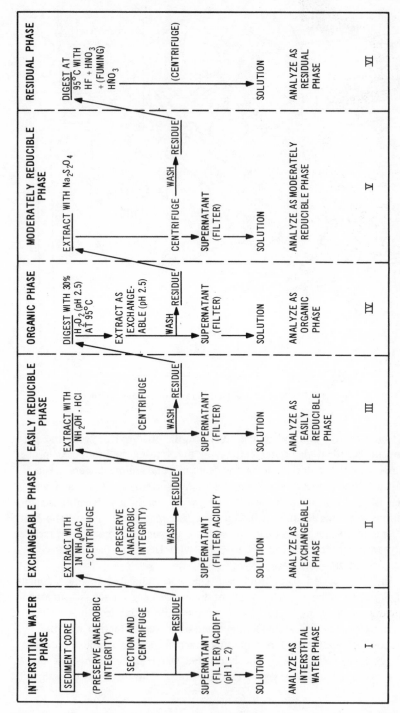

Figure 7.2 Selective extraction scheme for sediment characterization.[14]

the easily reducible phase, was obtained by extracting the sediment residue from the exchangeable phase with 0.1 M of hydroxylamine hydrochloride in 0.01 M of nitric acid. The organic fraction, digested at 95°C with 30% hydrogen peroxide at pH 2.5, was then extracted with 1 N ammonium acetate at pH 2.5. The residue from the organic fraction was extracted with a solution of sodium dithionite-sodium citrate and designated as the moderately reducible phase. The residue of this extract was subjected to a hydrofluoric acid-nitric acid-fuming nitric acid digest and designated the residual phase.

Within five days after sampling, the interstitial water and exchangeable phases of the sediment were extracted; the remaining sediment fractionation scheme was carried out at a later date. Duplicate extractions were performed for each 15-cm depth interval within a core for all phases with the exception of the interstitial water phase. These results were averaged to obtain the metal concentration associated with each operationally defined sediment extract from the core.

Dispersed particle-size distribution[15] and cation exchange capacity[16] were determined on wet sediment samples for each core that was chemically fractionated. The cation exchange capacity procedure was modified for centrifugation to include saturating wet sediment samples with ammonia, removing the excess ammonia with an isopropyl alcohol wash, and extraction of the sediments with a series of 2N KCl solutions.[17] Ammonium in the KCl extract was determined by nesselerization. Total carbon and organic carbon in the sediments were determined by dry combustion at 950°C,[18,19] with inorganic carbon assumed to be the difference between total and organic carbon. Total sulfides were determined on sediment samples frozen under nitrogen by a modification of the method published by Farber.[20,21]

The concentrations of iron, manganese, zinc and copper were determined in all fractions of the selective extraction scheme with a Perkin-Elmer Model 503 atomic absorption spectrophotometer equipped with a deuterium arc background corrector. Copper and zinc concentrations below the detection limit of the instrument were determined directly on the extracts by the method of standard additions using a Perkin-Elmer Model 2100 heated graphite atomizer.

In addition to an analysis of variance between sites and depths within sites, simple linear correlations were run to see what relationship, if any, existed among the trace metals in the chemical fractions.

RESULTS

Analysis of Variation

To test the validity of the selective extraction procedure, duplicate extractions were run upon every sample taken with the exception of the interstitial water, which was limited by sample size. The data were then statistically analyzed to give the coefficient of variation (CV) between duplicate extractions. These data are presented in Table 7.1 and are available for iron and manganese in all fractions but the interstitial water, where the CV given reflects the differences between cores in a site and laboratory error.

Table 7.1   Coefficient of Variation for Laboratory Replication of the Selective Extraction of Fe and Mn in Mobile Bay, Alabama

Extraction	Coefficient of Variation (%)	
	Fe	Mn
Interstitial water	30.4	21.1
Ammonium acetate	7.0	–
Hydroxylamine hydrochloride	8.8	9.7
Acidified H_2O_2	14.0	15.9
Citrate-dithionite	9.9	10.7
Residual	4.6	22.3

A check of the sum of all fractions for each core and depth within a site compared to concentrations obtained using a total digest (analytical total) showed that copper, manganese and iron were within 1.0, 3.9 and 33.3% of the total sediment concentration, respectively. The close agreement between replicates in all fractions and the complete recovery of copper and manganese indicated both precision and accuracy in the extraction procedures. The small manganese and copper error appeared to be due to a slight carryover between fractions; however, further experimentation indicated that most of the iron was lost in the wash following the citrate-dithionite extraction. Washing the citrate-dithionite extraction residue with additional citrate-dithionite solution produced no additional iron; the extraction residue with deionized water extracted additional iron. The reasons for this additional extraction with deionized water are not known at this time, but it is possibly due to the decomposition of sodium dithionite during the wash. This loss of iron in the citrate-dithionite wash did not seriously affect manganese and copper, which were present in small quantities in the citrate-dithionite extract relative to their abundance in other extracts. The iron and other trace metals lost after the citrate-dithionite extraction would have increased the trace metal content of the residual phase, but this problem can be rectified by dispensing with the wash after the extraction or by washing with additional citrate-dithionite solution.

The hydroxylamine hydrochloride extraction method of Chao[22] proved to be more effective than the method of Chester and Hughes[1] for chemically separating the oxides of manganese and iron. Chao's method extracted approximately 85% of manganese oxides while dissolving only 5% of the total iron present in Mobile Bay sediments, concurring with data obtained in other soils and sediments by Chao.[22]

Physical Sediment Characteristics

Table 7.2 details the physical and physicochemical characteristics of the Mobile Bay sediment cores. Particle size distribution showed the greatest change between sites in the > 50 μ-diameter fraction, which decreased sharply in site 2 as the <2 μ-diameter fraction increased in site 2. Cation exchange capacity averaged 44.5 meq/100 g and did not vary greatly either with site or depths within sites. Total organic carbon was significantly higher in site 2 but showed no significant change with depth within sites. Total sulfides were high, indicating high reducing conditions for both sites especially the 0- to 15-cm depth of site 2, where total sulfide was much higher than in other depths within the site. Site 1 contained less total sulfides at the sediment surface than contained in lower depths.

Chemical Properties of Water Samples

Table 7.3 presents some chemical and physicochemical properties of Mobile Bay waters at the study sites. No trends were apparent in trace metal concentrations near the sediment-water interface (8.0-m water depth), but manganese and copper concentrations were higher than those reported in Gulf waters near Alabama.[23] A salinity gradient was observed at site 1 along with reduced dissolved oxygen concentration near the water-sediment interface. Water column data were not obtained at site 2 because of an equipment malfunction.

Selective Extraction Results

Data presented in Figures 7.3-7.10 are the mean values of each depth segment for three cores per site. Tables 7.4-7.7 present the iron,

Table 7.2 Physical and Physicochemical Characteristics of Mobile Bay Sediments

Site	Depth (cm)	Particle Size Distribution (%)			Cation Exchange Capacity (meq/100g)	Total Organic Carbon (%)	Inorganic Carbon (%)	Total Sulfides (mg/g)	pH	Eh (mV)	Temp. (°C)
		<2μ	2-50μ	>50μ							
1	0-15	43.3	27.7	29.0	47.3	1.89	0.11	0.76	7.1	-225	17
	15-30	40.3	30.7	29.0	43.2	1.71	0.16	1.33	–	–	–
	30-45	36.0	29.0	35.0	43.1	1.95	0.10	0.86	–	–	–
	45-60	50.3	33.0	15.7	49.8	1.94	0.12	1.00	–	–	–
2	0-15	62.0	31.7	6.3	45.9	2.25	0.03	1.02	6.7	-185	17
	15-30	59.3	37.0	3.7	43.8	2.19	0.08	0.72	–	–	–
	30-45	58.7	36.7	4.7	42.3	2.28	0.03	0.57	–	–	–
	45-60	59.3	35.7	5.0	40.7	2.25	0.05	0.53	–	–	–

Table 7.3 Chemical and Physicochemical Properties of Mobile Bay Waters at the Sampling Sites

Site	Depth (m)	Temp. (°C)	Salinity (%)	Dissolved Oxygen (mg/l)	pH	Cu (μg/l)	Mn (μg/l)	Fe (μg/l)	Zn (μg/l)
1	0	13.7	1.0	10.8	5.9	–	–	–	–
	8.0	15.0	12.5	3.6	6.1	5	90	25	4.2
2	0	–	–	–	–	–	–	–	–
	8.0	–	–	–	–	2	130	35	8.0

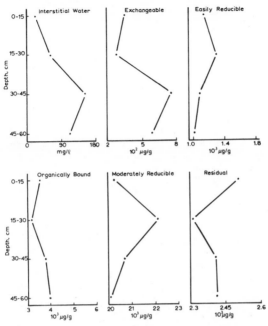

Figure 7.3   Distribution of iron in the chemical fractions of site 1.

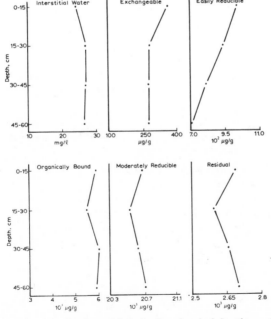

Figure 7.4   Distribution of iron in the chemical fractions of site 2.

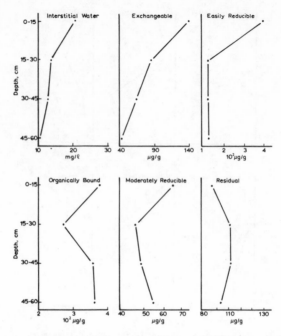

Figure 7.5 Distribution of manganese in the chemical fractions of site 1.

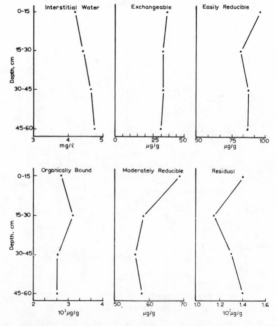

Figure 7.6 Distribution of manganese in the chemical fractions of site 2.

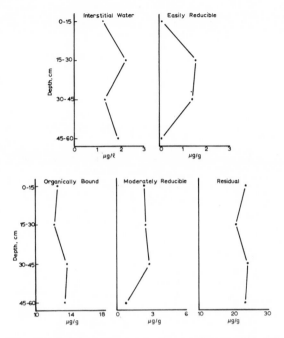

Figure 7.7 Distribution of copper in the chemical fractions of site 1.

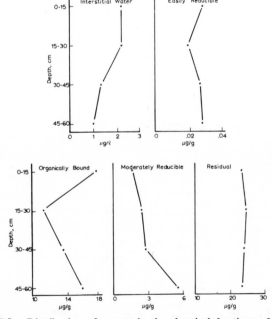

Figure 7.8 Distribution of copper in the chemical fractions of site 2.

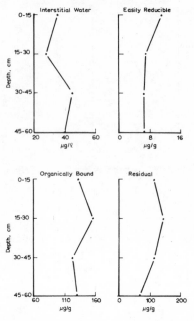

Figure 7.9   Distribution of zinc in the chemical fractions of site 1.

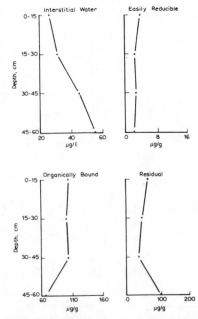

Figure 7.10   Distribution of zinc in the chemical fractions of site 2.

Table 7.4   Iron Content in Chemical Fractionation of Mobile Bay, Alabama, Sediments

| | Iron Content % of Total | | | | | | | |
| | Site 1–Depth, cm | | | | Site 2–Depth, cm | | | |
Chemical Fraction	0-15	15-30	30-45	45-60	0-15	15-30	30-45	45-60
Interstitial water	0.08	0.21	0.52	0.41	0.08	0.09	0.09	0.08
Exchangeable	1.27	0.89	2.57	2.08	0.89	0.93	0.90	0.89
Easily reducible	4.10	5.85	3.73	3.63	2.28	2.83	2.51	2.28
Organically bound	12.67	10.32	13.36	13.91	19.41	18.53	20.43	19.41
Moderately reducible	72.75	72.66	71.38	71.22	68.54	68.89	67.39	68.54
Residual	9.11	7.76	8.34	8.76	8.79	8.73	8.69	8.79
Sum (%)	2.80	3.04	2.90	2.81	3.07	2.97	3.06	3.07
Analytical total (%)	4.36	4.96	4.34	4.47	4.50	4.28	4.60	4.07

Table 7.5   Manganese Content in Chemical Fractionation of Mobile Bay, Alabama, Sediments

| | Manganese Content % of Total | | | | | | | |
| | Site 1–Depth, cm | | | | Site 2–Depth, cm | | | |
Chemical Fraction	0-15	15-30	30-45	45-60	0-15	15-30	30-45	45-60
Interstitial water	2.03	2.67	2.20	1.14	0.66	0.70	0.80	0.80
Exchangeable	12.90	13.52	8.86	4.75	7.22	7.07	7.19	6.59
Easily reducible	35.50	18.50	16.54	37.91	14.92	13.70	14.90	14.41
Organically bound	35.53	41.86	49.24	37.69	44.36	47.72	45.91	45.07
Moderately reducible	5.97	7.33	7.04	5.92	10.74	9.81	9.41	9.59
Residual	8.07	16.12	16.17	12.75	22.12	20.95	21.79	23.54
Sum (ppm)	1069	626	701	926	638	588	587	594
Analytical Total (ppm)	1074	795	696	685	599	555	586	522

Table 7.6 Copper Content in Chemical Fractionation of Mobile Bay, Alabama, Sediments

| | Copper Content % of Total | | | | | | | |
| | Site 1—Depth, cm | | | | Site 2—Depth, cm | | | |
Chemical Fraction[a]	0-15	15-30	30-45	45-60	0-15	15-30	30-45	45-60
Interstitial water (ppb)	1.21	2.26	1.40	1.90	2.24	2.27	1.33	1.00
Easily reducible	0.24	4.70	3.61	0.05	0.05	0.05	0.04	0.04
Organically bound	32.63	32.36	32.50	35.77	41.71	33.80	28.99	34.77
Moderately reducible	6.11	6.43	6.64	1.80	13.83	6.90	6.28	12.03
Residual	61.02	56.51	57.25	62.35	51.41	59.25	64.69	50.95
Sum (ppm)	37.8	37.0	41.6	37.1	42.8	40.0	37.9	44.7
Analytical total (ppm)	36.3	41.0	35.1	41.5	41.5	39.0	41.4	39.8

[a] No copper was extracted in the exchangeable fraction.

Table 7.7 Zinc Content in Chemical Fractionation of Mobile Bay, Alabama, Sediments

| | Zinc Content % of Total | | | | | | | |
| | Site 1—Depth, cm | | | | Site 2—Depth, cm | | | |
Chemical Fraction[a]	0-15	15-30	30-45	45-60	0-15	15-30	30-45	45-60
Interstitial water (ppb)	35.3	27.3	**45.3**	40.5	26.0	31.3	46.3	56.0
Easily reducible	4.15	2.25	2.98	2.98	2.25	1.71	1.81	1.28
Organically bound	52.04	51.13	52.22	65.37	58.97	67.76	72.18	38.03
Residual (by difference) (by difference)	43.79	46.60	45.23	31.64	38.76	30.51	26.00	60.68
Analytical total (ppm)	257	306	248	213	177	149	149	188

[a] No zinc was extracted in the exchangeable or moderately reducible fraction.

manganese, copper and zinc data, respectively, from Mobile Bay sediments as a percentage of the total extracted by the selective extraction procedure. The sum of the fractions is presented for each depth within a site along with the analytical total for comparison purposes.

Iron

The selective extraction of sediment iron in both sites (Figures 7.3 and 7.4) revealed that the greatest portion of iron was extracted by citrate-dithionite (moderately reducible phase). Such a response indicated that the majority of sediment iron existed as free iron oxides.[24] Additionally, the moderately reducible iron was rather uniform with no statistical difference between sites or depths within sites.

Organically bound iron (H_2O_2 extractable) was of secondary importance as a reservoir of sediment iron even though an unknown quantity of iron may have been contributed to this fraction by the oxidation of iron sulfides. Such a contribution to this and other fractions was indicated by a significant linear correlation (r=0.89**) between easily reducible iron and total sulfides, indicating that FeS or an oxidation product of FeS was making an important contribution to the iron content of the easily reducible fraction. Organically bound iron showed no change with depth in either site, but site 2 contained significantly greater organically bound iron. Residual iron bound in the crystal lattices of minerals was third in importance as a reservoir of sediment iron, but the total amount of residual iron was underestimated due to the loss in the wash after the citrate-dithionite extraction. As with organic iron, residual iron was higher in site 2 but showed no significant change with depth in either site. Hydroxylamine hydrochloride extractable iron (easily reducible) varied significantly between sites and with depth within sites with site 1 being highest. Different trends with depth were exhibited, but each followed the distribution of total sulfides within the site (Table 7.2).

Interstitial water and ammonium acetate extractable (exchangeable) iron showed much the same trends within a site except in the surface depth of site 2. Site 1 was highest in interstitial water iron and showed a gradual increase with depth until the 45- to 60-cm depth, where a decrease occurred. Site 2 showed no significant change in interstitial water iron with depth. Interstitial water iron was more concentrated than in the overlying water and was also much higher in both sites than interstitial water concentrations reported in deep-sea sediments[2,25] or in shallow water marine sediments.[26]

Manganese

The selective chemical extraction data for manganese indicated greater differences in fractional distribution for manganese than for iron (Figures 7.5 and 7.6). The largest portion of sediment manganese was found in the organically bound fraction. However, as with iron, the most abundant fraction was the most uniform, showing no significant difference between sites or depths within sites. Easily reducible manganese was the second largest reservoir of sediment manganese in site 1, but this phase contained less manganese than the residual phase in site 2. Manganese carbonates not extracted by the ammonium acetate could have contributed to the easily reducible phase fraction, but the amount of carbonates in the sediment was very small. Manganese was found in small quantitites in the moderately reducible phase; there was no significant difference between sites, but the top depth in each site had the highest manganese content.

Interstitial water manganese was enriched in both sites relative to the overlying water. Values of interstitial water manganese were higher for both sites than reported values for deep-sea sediments,[2,25,27] more shallow marine sediments[26] and in Lake Ontario.[28] Interstitial water manganese showed decreasing content with depth in site 1 but increased slightly with depth in site 2. Exchangeable manganese showed basically the same trends with depth as in interstitial water manganese in both sites.

Copper

The sediment copper appeared to be concentrated in the organic and residual phases of sediment extracts (Figures 7.7 and 7.8). Very little copper was extracted by hydroxylamine hydrochloride or citrate-dithionite, and no correlation existed between copper and the manganese or iron extracted in these two fractions. Copper was depleted in the interstitial water compared to the overlying water and was also less than that observed in shallow water and deep-sea areas.[26,29] No copper was extracted by the ammonium acetate extractant. The background copper concentrations of 60 ppb decreased to approximately 3 ppb after extraction. Twenty grams of sediment (dry weight basis) extracted with 100 ml of the ammonium acetate solution produced 0.4 μg of copper sorbed per gram of sediment.

More organically bound copper was present in site 2, indicative of the higher organic carbon content of these sediments. Residual copper showed little change between sites or with depth within sites. The copper fractionation data agreed well with the findings of Presley, et al.[2]

in Saanich Inlet sediments and Bruland et al.[4] in the southern California coastal zone. All found the majority of sediment copper in the organically bound and residual phases.

Zinc

Sediment zinc was concentrated in the organically bound and residual phases of the sediment extracts (Figures 7.9 and 7.10). No zinc was extracted by the citrate-dithionite reagent and zinc also behaved similarly to copper in the exchangeable phase where extraction served to decrease the slight zinc contamination in the reagent. Interstitial water zinc was enriched relative to the overlying water and was of the same order of magnitude observed by other workers[2,26,30] in the interstitial waters of marine sediments. Interstitial water zinc showed no significant differences between sites, but site 2 exhibited a steady increase with depth not observed in site 1. More zinc was extracted in the easily reducible phase in site 1 with site 1 also exhibiting a highly significant positive correlation ($r=0.89**$) between easily reducible zinc and manganese.

Organically bound zinc showed significant changes between sites, but no significant change was seen with depth within sites. Residual zinc showed no significant changes between sites or with depth within sites. Residual zinc concentrations given are the difference between the analytical total and the sum of the remaining zinc fractions. Contamination of the residual phase by zinc carryover from the citrate-dithionite reagent necessitated this step. The zinc fractionation data agreed well with the data of Presley et al.[2] who fractionated shallow cores from Saanich Inlet, British Columbia.

DISCUSSION

The selective dissolution scheme used in this study provides a more detailed distribution pattern of trace metals than other reported methods.[1,2,31] It must be borne in mind, however, that the distribution of trace metals in any chemically extracted fraction is defined by the method of extraction rather than fundamental properties of the system.[7]

Based on the Mobile Bay data, the selective extraction techniques of Engler, et al.[14] appear to have adequate precision and accuracy for characterizing the distribution of iron and manganese in sediments. Replicate extractions analyzed for copper and zinc agreed as closely as iron and manganese in all fractions.

The highly significant correlation between total sulfides and iron extracted by acidified hydroxylamine hydrochloride could be due to a number of factors, such as iron sulfide dissolution, or iron oxides freshly formed from FeS oxidation may have been more susceptible to reduction than older crystalline iron oxides. Freshly formed oxides could have been present since no special precautions were taken after the ammonium acetate extraction to prevent air oxidation of the residue; however, the container was sealed and stored at 4°C until extracted with the acidified hydroxylamine hydrochloride. Ponnamperuma[11] reported that Asami[32] observed an increased percentage of reducible iron oxides associated with a lower degree of iron oxide crystallinity in reduced soil samples. Trace metals associated with carbonates would also contribute to the trace metal content of the acidified hydroxylamine hydrochloride extract, especially in calcareous sediments, but these sediments contained very little inorganic carbon. Additionally, manganese carbonate is appreciably soluble in ammonium acetate and could therefore be expected to contribute to the manganese concentrations found in the exchangeable phase.

The high concentrations of interstitial water iron and manganese found in Mobile Bay sediments relative to the concentrations found in deep-sea sediments may have been due to a number of environmental factors, such as iron and manganese transformation to a more soluble state as the sediment becomes more reduced, with a resultant concentration change in the interstitial water. Ponnamperuma et al.[33] considered the concentration of water-soluble iron in most submerged soils to be controlled by the effects of redox potential and pH on the $Fe(OH)_3$-Fe^{2+} system with a more stable concentration regulated by the $Fe_3(OH)_8$-Fe^{2+} equilibrium.

In marine sediments, the precipitation of iron as its sulfide will also assume an important role in controlling the concentration of iron and other heavy metals in the interstitial water. Duchart et al.[26] attributed lower concentrations of iron in interstitial water to the greater ease of formation and stability of FeS compared to MnS. Despite the high total sulfide content of the Mobile Bay sediments, iron was more concentrated in the interstitial water than manganese. The high interstitial water concentrations could be indicative of organic chelation, the control of Fe concentration in the interstitial water by some unspecified solid phase other than FeS,[30] or the presence of metastable metal-sulfide polymers.[34] Also the sediment may have been depleted of sulfate, precluding further H_2S formation and resulting in increased interstitial water iron. Serruya et al.[35] found sulfates diffusing from lake waters into sediments to be rapidly reduced in the surface sediment layers with the absence or low values of H_2S in deeper sediments causing the extension of the field of maximum solubility of iron. No odor of H_2S was evident in these Mobile Bay sediments, indicating that very low concentrations of H_2S were present.

Duchart et al.[26] attributed the enrichment of trace metals in pore waters to the presence of dispersed oxides of iron and manganese at the surface of recent sediments and their ability to scavenge trace metals. Burial and subsequent reduction of this oxide material were postulated as a mechanism of trace metal accumulation. The concentrations of iron and manganese in Mobile Bay interstitial waters were much higher than those noted by Duchart et al.[26] in shallow marine sediments, possibly indicative of the burial and reduction of a greater amount of hydrous iron and manganese oxides in the Mobile Bay sediments than in recent sediments. Iron in the interstitial water increased sharply with depth in site 1 following trends observed by Brooks et al.,[30] who attributed increases in interstitial water iron with depth to solubilization by the lowering of Eh and pH with depth in reducing sediments. Duchart et al.[26] attributed decreased interstitial water manganese with depth in reduced sediments to possible precipitation as manganese carbonates.

Evidence of manganese mobilization and subsequent migration to a surface oxidized zone existed in site 1. In sediments with reducing conditions underlying a more oxidized surface sediment zone, manganese will migrate in the pore waters due to a concentration gradient toward the surface where oxidation and precipitation may occur, resulting in a net upward migration.[26,27,36] Gorham and Swaine[37] attributed iron and manganese enrichment in the oxidate crusts of lake sediments to migration and precipitation. In Mobile Bay sediments, total manganese was highest at the sediment surface and decreased with depth. Easily reducible manganese was four times higher in the 0- to 15-cm sediment segment than in any of the adjacent sediment segments. It is possible that manganese oxides were concentrated in a thin oxidized surface layer that was diluted when blended with the 15-cm segment of core sampled.

Bonatti et al.[38] noted that in deep-sea sediments, depletion of solid phase manganese should be paralleled by an increase of dissolved manganese in the interstitial water. This increase in interstitial water manganese with depth was not noted in these estuarine sediments, which may be indicative of manganese precipitation from the overlying water resulting in increased manganese content in the surface sediment. It is thought that migration to the surface oxidized zone from the reduced sediment is controlled largely by ionic and/or molecular diffusion resulting mainly from the interstitial waters having a higher content of a particular element than the overlying water.[39] Mobilization and migration in site 1 would have to be a rapid process with solubilization and migration occurring soon after burial. Migration to the surface oxidized zone from greater depths is inhibited by the lack of a concentration gradient of interstitial water manganese from greater depths to the surface. The sediment deposited in Arlington Channel is of very recent origin, for the channel is dredged every two years to maintain a depth of 8.3 meters; the last dredging occurred in 1972. Mobilization of reducible manganese into the interstitial and exchangeable phases has been shown to occur rapidly with decreased redox potential at a natural pH.[9] Additionally, sufficient oxygen was in solution at the sediment-water interface to precipitate iron and manganese at the sediment surface;[40,41] however, manganese migration to a surface oxidized zone can only be postulated for this sediment.

The concentration gradient observed with depth for interstitial water iron in site 1 could result in the migration of iron to a surface oxidized zone. However, no change in total iron concentration was found with depth in either site, but small amounts of iron accumulation could have been masked by the large total iron content of the sediment. Lack of migration would be supported by the findings of Bonatti et al.,[38] who concluded that iron, zinc and copper do not migrate significantly in reduced sediments due to immobilization as sulfides. This appeared to hold true for copper, iron and zinc in Mobile Bay sediments, which showed no signs of surficial accumulation; however, the upward migration of iron was possible in site 1.

Site 2 displayed no changes with depth in any fraction for manganese and copper, and only the easily reducible fraction of iron and zinc in the interstitial water showed any significant change with depth. This lack of change with depth suggests a much more homogeneous sediment profile than found in site 1. Duchart et al.[26] attributed the lack of change with depth of interstitial water iron and manganese to a completely anoxic vertical sediment profile. This may be the case in site 2, but the only evidence to support this theory is the high total sulfide content of the surface layer in site 2, possibly indicating that conditions were very reducing at the sediment surface with a resultant anoxic sediment-water interface.

Moderately reducible manganese was slightly higher in the surface depth of both sites, giving further evidence of mobilization in site 1 and possibly indicating manganese mobilization and migration in site 2 followed by occlusion in the more insoluble iron oxides.

The decreased total and easily reducible manganese content in site 2 was indicative of a decrease in the manganese oxide content of the sediment. There was no significant change between sites in organically bound and moderately reducible manganese and little change in residual manganese. The decreased easily reducible manganese in site 2 was apparently not a result of transformation to another manganese phase or fraction.

The decreased manganese could have been lost to the overlying water if the sediment-water interface were anoxic[40,41] or less manganese could have been initially present in that sediment. The silt and sand ($> 50\,\mu$) fraction constituted less than 5% of the sediment at site 2, while silt and sand constituted 27% of site 1 sediment. The bulk of the manganese oxides present in site 1 was probably associated with this fraction of sediment, while site 2 had less manganese and less silt and clay. Robinson[42] found the silt fraction of a wide range of soils to be highest in manganese, the sand fraction to contain a significant amount of manganese, and little manganese occurring in the clay ($< 2\,\mu$) fraction.

Increased organically bound iron and copper in site 2 was associated with the increased organic carbon and may be indicative of a sink for trace metal contamination from the waterway. Cooper and Harris[43] found the distribution of metals as organic complexes in sediment to be very sensitive to environmental factors such as contamination. Presley et al.[2] concluded that diagenesis within sediment in Saanich Inlet, British Columbia, appeared to be time dependent. They found mobilization of copper, iron and zinc from the insoluble silicate phase to the peroxide-soluble phase, suggesting organic complexing. No evidence of this mechanism was apparent in Mobile Bay sediments, but the time frame for these processes was possibly too long to have caused the enrichment of the organically bound phase.

Both sites showed more copper associated with the moderately reducible phase than with the easily reducible phase. Goldberg[5] found the copper content of deep-sea sediments linearly related to the manganese content and explained their relationship in terms of the scavenging ability of manganese oxides. However, greater concentrations of copper in these Mobile Bay sediments were associated with the iron oxide fraction of the sediment, indicating a possible sorption mechanism involving iron oxides. Collins[44] attributed the loss of copper in a mine tailings stream to sorption by iron hydroxides, which may also be a mechanism of copper sorption in these sediments. McLaren and Crawford[45] found the majority of copper in soils organically bound and in clay lattice structures. Their data indicated that manganese has a greater effect than iron on the amount of copper occluded. However, since copper associated with the oxide phase is a function of both initial sorption and retention, it was proposed that manganese scavenges the copper and that the iron, due to its greater abundance, is responsible for the actual occlusion of copper in the oxides. We must also consider the possibility that copper released by the acidified hydroxylamine hydrochloride extract was sorbed by some other sediment phase during the course of the extraction procedure. As indicated by the sorption of copper and zinc from the ammonium acetate extractant (exchangeable phase), some sorption mechanism was at work in these Mobile Bay sediments, possibly the organic sediment phase for copper.

The association of zinc with manganese in the easily reducible phase agrees with the sorption mechanism for zinc of Krauskopf;[6] however, the sorption mechanism of McLaren and Crawford[45] for copper may also hold true for zinc. Zinc was not found in the moderately reducible phase; however, copper was found only in small quantities.

Zinc extracted in the peroxide digest was not correlated with total organic content of the sediments, suggesting that some forms of inorganic zinc are also contributing to this fraction. Piper[31] found that zinc not solubilized by a 0.1 M HCl leach was negatively correlated with sediment organic carbon. However, Presley et al.[2] analyzed plankton from Saanich Inlet, British Columbia, and found that zinc should be the metal most closely related to organic matter; yet analysis of sediment organically bound zinc showed no relation to the organic carbon content of the sediment. They postulated that zinc may be more readily released into the water column during biological degradation than other metals. Cooper and Harris[43] found copper and zinc associated with asphalts and humic acids in the sediments studied, yet lower concentrations of zinc were found in the total organic fraction after an acid leach when compared to the sum of the soluble organic fractions. Loss of zinc in the total organic fraction was attributed to acid hydrolysis.

CONCLUSIONS

The selective extraction scheme studied can determine the different geochemical fractions in a sediment, with excellent mass balance for copper and manganese. The most serious problems encountered were the zinc contamination of the sodium dithionite and the loss of iron when washing the citrate-dithionite extraction residue. However, the various chemical extractants and interstitial water samples within sites had small variation between replicates.

The heavy metals investigated, with the exception of copper, were enriched in the interstitial water relative to the overlying water. Iron and manganese concentrations were exceptionally high in the interstitial water, despite the high total sulfide content of the sediments. Metal profiles in the interstitial water revealed conditions potentially suitable for surficial accumulation of iron and zinc; however, no migration to surface layers could be conclusively shown, even in the case of manganese, which was enriched in the surface sediment of site 1.

The majority of sediment iron was found in the moderately reducible phase, with the organically bound fraction of secondary importance. However, residual iron was underestimated due to loss in the wash following the moderately reducible extraction. Organically bound manganese was the largest reservoir of sediment manganese; easily reducible or residual manganese were the next largest reservoirs, depending upon the site under consideration. Copper and zinc were concentrated in the organic and residual phases of sediment extracts, and copper was more uniformly distributed among sites than iron, manganese and zinc.

The sorption of copper and zinc from the 1 N ammonium acetate extractant illustrates the great value of sediments as a sink for some trace metals.

ACKNOWLEDGMENTS

The authors wish to thank Drs. P. G. Hunt and R. H. Plumb, Jr. for their many comments and suggestions during the preparation of this manuscript. This research was funded under the Dredged Material Research Program of the U.S. Army Corps of Engineers.

REFERENCES

1. Chester, R. and M. J. Hughes. "A Chemical Technique for the Separation of Ferro-Manganese Minerals, Carbonate Minerals and Adsorbed Trace Metals from Pelagic Sediments," *Chem. Geol.* 2, 249 (1967).
2. Presley, B. J., Y. Kolodny, A. Nissenbaum and I. R. Kaplan. "Early Diagenesis in a Reducing Fjord, Saanich Inlet, British Columbia — II. Trace Element Distribution in Interstitial Water and Sediment," *Geochim. Cosmochim. Acta* 36, 1073 (1972).
3. Nissenbaum, A. "Distribution of Several Metals in Chemical Fractions of Sediment Core from the Sea of Okhotsk," *Israel J. Earth Sci.* 21, 143 (1972).
4. Bruland, K. W., K. Bertine, M. Koide and E. D. Goldbert. "History of Metal Pollution in Southern California Coastal Zone," *Environ. Sci. Technol.* 8(5), 425 (1974).
5. Goldberg, E. D. "Marine Geochemistry 1. Chemical Scavengers of the Sea," *J. Geol.* 62, 249 (1954).
6. Krauskopf, K. B. "Factors Controlling the Concentrations of Thirteen Rare Metals in Sea Water," *Geochim. Cosmochim. Acta* 9, 1 (1956).
7. Jenne, E. A. "Controls on Mn, Fe, Co, Ni, Cu, and Zn Concentrations in Soils and Water: The Significant Role of Hydrous Mn and Fe Oxides," in *Trace Inorganics in Water, Advances in Chemistry Series*, R. F. Gould, ed. (1968), pp. 337-387.

8. Gibbs, R. J. "Mechanisms of Trace Metal Transport in Rivers," *Science* **180**, 71 (1973).

9. Gotoh, S. and W. H. Patrick, Jr. "Transformation of Manganese in a Waterlogged Soil as Affected by Redox Potential and pH," *Soil Sci. Soc. Amer. Proc.* **36**, 738 (1972).

10. Gotoh, S. and W. H. Patrick, Jr. "Transformation of Iron in a Waterlogged Soil as Influenced by Redox Potential and pH," *Soil Sci. Soc. Amer. Proc.* **38**, 66 (1974).

11. Ponnamperuma, F. N. "The Chemistry of Submerged Soils," *Adv. Agron.* **24**, 29 (1972).

12. Patrick, W. H., Jr. "Extractable Iron and Phosphorus in Submerged Soil at Controlled Redox Potential," *Internat. Cong. Soil Sci. Trans.* 8th (Bucharest, Romania) **IV**, 605 (1964).

13. Sholkovitz, E. "Interstitial Water Chemistry of the Santa Barbara Basin Sediments," *Geochim. Cosmochim. Acta* **37**, 2043 (1973).

14. Engler, R. M., J. M. Brannon, J. R. Rose and G. N. Bigham. "A Practical Selective Extraction Procedure for Sediment Characterization," presented at Symposium on Chemistry of Marine Sediments, National American Chemical Society Meeting, Atlantic City, New Jersey, August 12, 1974.

15. Patrick, W. H., Jr. "Modification of Method of Particle Size Analysis," *Soil Sci. Soc. Amer. Proc.* **22**, 366 (1958).

16. Jackson, M. L. *Soil Chemical Analysis.* (Englewood Cliffs, New Jersey: Prentice Hall, Inc., 1958).

17. Bremner, J. M. "Inorganic Forms of Nitrogen," in *Methods of Soil Analysis, Part 2. Agron.* **9**, 1179 (1965).

18. Allison, L. E., W. Boller and C. D. Moodie. "Total Carbon," in *Methods of Soil Analysis, Part II. Agron.* **9**, 1346 (1965).

19. Allison, L. E. "Organic Carbon," in *Methods of Soil Analysis, Part II. Agron.* **9**, 1367 (1965).

20. Farber, L. *Standard Methods for the Examination of Water and Wastewater,* 11th ed. (New York: American Public Health Association, Inc., 1960).

21. Connell, W. E. "The Reduction of Sulfate to Sulfide Under Anaerobic Soil Conditions," M.S. Thesis, Louisiana State University, Baton Rouge (1966).

22. Chao, T. T. "Selective Dissolution of Manganese Oxides from Soils and Sediments with Acidified Hydroxylamine Hydrochloride," *Soil Sci. Soc. Amer. Proc.* **36**, 764 (1972).

23. Slowey, J. F. and D. W. Hood. "Copper, Manganese and Zinc Concentration in Gulf of Mexico Waters," *Geochim. Cosmochim. Acta* **35**, 121 (1971).

24. Holmgren, G. G. S. "A Rapid Citrate-Dithionite Iron Procedure," *Soil Sci. Soc. Amer. Proc.* **31**, 210 (1967).

25. Presley, B. J., R. R. Brooks and I. R. Kaplan. "Manganese and Related Elements in the Interstitial Water of Marine Sediments," *Science* **158**, 906 (1967).

26. Duchart, P., S. E. Calvert and N. B. Price. "Distribution of Trace Metals in the Pore Waters of Shallow Water Marine Sediments," *Limnol. Oceanog.* **18**, 605 (1973).

27. Bischoff, J. L. and T. L. Ku. "Pore Fluids of Recent Marine Sediments; II., Anoxic Sediments of 35° to 45° N Gibraltar to Mid-Atlantic Ridge," *J. Sed. Petrol.* **41**, 1008 (1971).

28. Weiler, R. R. "The Interstitial Water Composition in the Sediments of the Great Lakes–I Western Lake Ontario," *Limnol. Oceanog.* **18**, 918 (1973).

29. Yemel'Yanov, Ye. M. and N. B. Vlasenko. "Concentrations of Dissolved Forms of Fe, Mn and Cu in Marine Pore Waters of the Atlantic Basin," *Geochem. Internat.* **9**, 855 (1972).

30. Brooks, R. R., B. J. Presley and I. R. Kaplan. "Trace Elements in the Interstitial Waters of Marine Sediments," *Geochim. Cosmochim. Acta* **32**, 397 (1968).

31. Piper, P. Z. "The Distribution of Co, Cr, Cu, Fe, Mn, Ni and Zn in Framvaren, a Norwegian Anoxic Fjord," *Geochim. Cosmochim. Acta* **35**, 531 (1971).

32. Asami, T. *Nippon Dojo - Hiryogakn Zasshi* **41**, 56 (1970).

33. Ponnamperuma, F. N., E. M. Tianco and T. Loy. "Redox Equilibria in Flooded Soils: 1. The Iron Hydroxide Systems," *Soil Sci.* **103**, 374 (1967).

34. Chen, K. Y. and J. C. Morris. "Kinetics of Oxidation of Aqueous Sulfide by O_2," *Environ. Sci. Technol.* **6**(6), 529 (1972).

35. Serruya, C., M. Edelstein, U. Pollingher and S. Serruya. "Lake Kinneret Sediments: Nutrient Composition of the Pore Water and Mud Water Exchanges," *Limnol. Oceanog.* **19**, 489 (1974).

36. Li, Y., J. L. Bischoff and G. Mathieu. "The Migration of Manganese in the Arctic Basin Sediment," *Earth Plant Sci. Letters* **7**, 265 (1969).

37. Gorham, E. and D. J. Swaine. "The Influence of Oxidizing and Reducing Conditions upon the Distribution of Some Elements in Lake Sediments," *Limnol. Oceanog.* **10**, 268 (1965).

38. Bonatti, E., D. E. Fisher, O. Joensuu and H. S. Rydell. "Post-Depositional Mobility of some Transition Elements, Phosphorus, Uranium and Thorium in Deep Sea Sediments," *Geochim. Cosmochim. Acta* **35**, 189 (1971).

39. Lynn, D. C. and E. Bonatti. "Mobility of Manganese in Diagenesis of of Deep Sea Sediments," *Mar. Geol.* **3**, 457 (1965).

40. Mortimer, C. H. "The Exchange of Dissolved Substances Between Mud and Water in Lakes," *J. Ecol.* **29**, 280 (1941).

41. Mortimer, C. H. "The Exchange of Dissolved Substances Between Mud and Water in Lakes," *J. Ecol.* **30**, 147 (1942).

42. Robinson, W. O. "Detection and Significance of Manganese Dioxide in the Soil," *Soil Sci.* **27**, 335 (1929).

43. Cooper, R. S. and R. C. Harris. "Heavy Metals in Organic Phases of River and Estuarine Sediment," *Mar. Poll. Bull.* **5**, 24 (1974).

44. Collins, B. I. "The Concentration Control of Soluble Copper in a Mine Tailings Stream," *Geochim. Cosmochim. Acta* **37**, 69 (1973).

45. McLaren, R. G. and D. V. Crawford. "Studies on Soil Copper I. The Fractionation of Copper in Soils," *J. Soil Sci.* **24**, 172 (1973).

8

RECENT DEPOSITION OF LEAD OFF THE COAST OF SOUTHERN CALIFORNIA

Steven M. Murray and Teh-Lung Ku

Department of Geological Sciences
University of Southern California
Los Angeles, California 90007

INTRODUCTION

The effect of man-generated metals on the marine environment has been the subject of concern in recent years. In the Southern California Bight, distinct enrichment of trace metals above their natural background has been noted in surface sediments near major sewage outfalls.[1] The amount of lead introduced into the oceans and its mode of transport, either direct land runoff or atmospheric burden, are vital questions. To accurately assess the flux of lead into the marine sediments, a precise knowledge of the modern sedimentation rate is essential. Sediment cores from three offshore areas of southern California were examined to establish the flux of lead to sediments and to estimate the importance of atmospheric transport (Figure 8.1).

These three areas lie in a line progressing away from the coast. For each core the chronology was determined by the Pb-210 method, and the lead concentrations were established.

METHODS

A box core was used to collect sediments from the sea floor in the three areas sampled, and extreme care was taken to ensure surface retention. Immediately after retrieval the cores were subsampled at 0.5-1 cm intervals and refrigerated until processed in the laboratory.

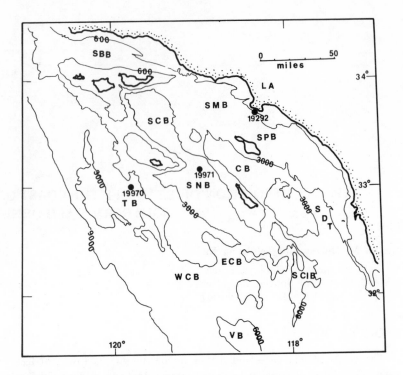

Figure 8.1 Core locations in the Southern California Borderland: SBB (Santa Barbara Basin); SMB (Santa Monica Basin), SCB (Santa Cruz Basin), TB (Tanner Basin), CB (Santa Catalina Basin), SNB (San Nicolas Basin), SPB (San Pedro Basin), SDT (San Diego Trough), ECB (East Cortes Basin), WCB (West Cortes Basin), SCIB (San Clemente Basin), VB (Velero Basin).

Pb-210 was measured utilizing the technique of Koide *et al.*[2] but with slight modifications. Approximately one gram of sediment was dried at 400°C and 40 mg of divalent lead was added as carrier. The sediment was then leached with hot 6 *N* HCl. The leachate was dried, dissolved in 1.5 HCl and passed through an anion exchange column (BIORAD AG1-X8). Lead was then eluted with deionized water and precipitated as a sulfate at a pH between 2 and 3. The lead sulfate was placed in a plastic tray and counted after 30 days. The time delay allows Bi-210 ($t_{1/2}$ = 5 days) to grow into equilibrium with Pb-210. The Bi-210 is counted because of its more energetic betas as compared with those of Pb-210 (1.2 MeV *vs* 0.02 MeV). The beta counter has a background of 8 counts per hour and an efficiency for Bi-210 of 27%.

Lead analysis was made by atomic absorption spectrophotometry of the totally dissolved sample. A deuterium background corrector was used with a precision of about 10%.

RESULTS

Pb-210 content of the sediments in these three cores is listed in Table 8.1. The unsupported Pb-210 is calculated by subtracting the Ra-226 supported Pb-210 from the measured total Pb-210. In the old portion

Table 8.1   Total Pb-210 Contents (dpm/g) of Sediment Cores

Depth (cm)	Core 19970 (Tanner Basin)	Core 19971 (San Nicolas Basin)	Depth (cm)	Core 19292 (San Pedro Shelf)
0-0.5	34.6	33.8	0-1	38.8
0.5-1	44.0	48.6	1-2	30.4
1-1.5	27.5	55.1	2-3	30.0
1.5-2	29.9	44.0	4-5	32.5
2-2.5	22.6	35.5	6-7	26.2
2.5-3	21.1	37.0	8-9	27.1
3-3.5	18.7	−	10-11	16.8
3.5-4	15.8	35.5	12-13	16.0
4-4.5	13.3	28.0	14-15	13.7
4.5-5	15.9	32.2	16-17	11.0
5-6	14.1	27.5	18-19	6.2
6-7	12.6	20.6	22-23	5.2
10-11	−	10.9	26-27	6.6
14-15	11.9	11.8	32-33	3.3
20-21	13.1	9.8	38-39	2.5

of the core the Pb-210 reaches a constant value, representing the Ra-226-supported component. To eliminate the sediment compaction factor, the sedimentation rates are converted into sediment flux values in the units $mg/cm^2/yr$. Table 8.2 summarizes the sediment flux averaged over the dated intervals including the estimated density of the solid sediments.[3]

The Pb concentrations in these three cores are shown in Table 8.3. The lead concentration in the 1900 level of the sediment represents the natural flux of lead to the sediment. By removing the natural flux from the recent flux the anthropogenic or man-generated flux may be estimated, as in Table 8.3.

DISCUSSION

The Pb-210 data demonstrates the applicability of the Pb-210 method to date sediments in the outer as well as the inner basins off the coast of

Table 8.2   Data on Sediment Flux

Core	Rate[a] (mm/yr)	H_2O-Free Bulk Density[b] (%)	Sediment Flux (mg/cm^2/yr)
San Pedro Shelf Core 19292	2.4	0.47	103
San Nicolas Basin Core 19971	1.3	0.44	57
Tanner Basin Core 19970	0.5	0.39	20

[a]Calculated from unsupported Pb^{210}.

[b]Solid density assumed 2.65 g/cm^3.

Table 8.3   Lead Concentration and Fluxes

Lead Concentration (ppm)				
Depth	Core 19970	Core 19971	Depth	Core 19292
0-0.5	19	19	0-1	40
0.5-1	20	13	1-2	20
1-1.5		13	2-3	23
1.5-2	16	19	3-4	28
2-2.5		14	4-5	27
2.5-3		11	7-8	26
3-3.5	15		8-9	27
3.5-4	13	14	10-11	17
4-4.5		15	14-15	13
4.5-5	12		16-17	16
5-6	16		18-19	12
6-7		10		
10-11	16			
14-15	9	12		

Lead Flux (μg/cm^2/yr)				
Anthropogenic	0.08	0.23	0.23	1.70

southern California. The anthropogenic flux of Pb in the San Pedro area (core 19292) is significantly above the limits of measurement errors. This observation is similar to those made by Bruland et al.[4] for three inner basins. In the two outer basins the anthropogenic flux of lead is much smaller, but also above the analytical error. The anthropogenic source of Pb is primarily auto emission.[5,6] Thus Pb is introduced into the atmosphere and is a good metal to establish the relative importance of atmospheric transport. Both Chow et al.[5] and Bruland et al.[4] stated that if atmospheric transport were dominant, the fluxes of a given element in the outer basins would be similar to those in the inner basins. If direct land runoff were the predominant mode of transport, the outer basin fluxes would be smaller. Upon examining the data presented here and utilizing their logic, one can say that the atmospheric transport and rainout over the depositional site is not the primary mode of transportation. As a limiting case, the anthropogenic flux of lead to the outer basins is about 10% of that to the inner basins (Table 8.3).

CONCLUSION

The Pb-210 geochronological technique is a valid dating technique for sediments extending the breadth of the Southern California shelf. The anthropogenic fluxes of Pb indicate that (1) atmospheric transport is not dominant and direct land discharge is dominant; or (2) the atmospheric transport produces a gradient over the 100-mile distance to the Tanner Basin.

ACKNOWLEDGMENTS

The authors wish to thank Dr. Chen of the University of Southern California for the use of his atomic absorbtion spectrophotometer. This research was supported by the Oceanography section, National Science Foundation, NSF Grant DES72-01557.

REFERENCES

1. Galloway, J. "Man's Alteration of the Natural Geochemical Cycle of Selected Trace Metals," Ph.D. dissertation, University of California, San Diego (1972).
2. Koide, M., A. Soutar and E. Goldberg. "Marine Geochronology with Pb-210," Earth Planet. Sci. Letters 14, 442 (1972).
3. Emery, K. O. The Sea Off Southern California (New York: John Wiley and Sons, 1969), p. 359.
4. Bruland, K., K. Bertine, M. Koide and E. Goldberg. "History of Metal Pollution in Southern California Coastal Zone," Environ. Sci. Technol. 8, 425 (1974).

5. Chow, T., K. Bruland, K. Bertine, A. Soutar, M. Koide and E. Goldberg. "Pb Pollution: Records in Southern California Coastal Sediments," *Science* **181**, 551 (1973).
6. Friedlander, S. "Chemical Element Balances and Identification of Air Pollution Sources," *Environ. Sci. Technol.* **7**, 235 (1973).
7. Southern California Coastal Water Research Project (SCCWRP). *The Ecology of the Southern California Bight.* Vol. 1 (1972), p. 531.

9

FATE OF PESTICIDES IN BOTTOM SEDIMENTS DURING DREDGING AND DISPOSAL CYCLE

Raymond J. Krizek

Department of Civil Engineering
The Technological Institute
Northwestern University
Evanston, Illinois 60201

Leo A. Raphaelian

Analytical Services Laboratory
Department of Chemistry
Northwestern University
Evanston, Illinois 60201

INTRODUCTION

Recent concerns for protection of the environment have dictated that many of the polluted bottom sediments dredged from our nation's harbors and waterways be deposited within diked containment areas. In the case of a hopper dredge, which is frequently used for this purpose, these sediments are mixed with large quantities of ambient river water and pumped into the disposal area in the form of a slurry with about 15% solids content. Then the solids settle out of suspension, and the excess water usually flows over a weir back into the harbor or river; on rare occasions a filtering system of some type is used in place of the weir. Accordingly, the effectiveness of this dredging and disposal operation depends largely on the extent to which the pollutants are associated with the solids retained within the containment area.

Many previous studies have indicated that polluted dredgings may contain a variety of natural wastes, fertilizers, pesticides, detergents, metals, oil, grease and many other by-products from industrial processes. Because of the high degree of pollution in many cases, microorganisms that normally degrade the pollutants and purify the water are unable to survive; moreover, even if they were able to survive, some industrial chemicals, particularly many of the pesticides, are intractable. This study was directed toward evaluating the extent to which various pesticides are associated with the solid portion of the dredgings and advancing some possible correlations and explanations for the observed behavior.

SAMPLES TESTED

Four samples of dredged materials from three Great Lakes harbors were taken while they were in the process of being deposited in diked containment areas. These samples are designated by a letter (C indicates a slurry with approximately 10-20% solids content; D represents a mud with a solids content of 25% or more), a chronological number, and two letters representing the city and state from which they were obtained (CO is Cleveland, Ohio; DM is Detroit, Michigan; and TO is Toledo, Ohio). The samples were taken using techniques described by Hummel and Krizek,[1] and some of the problems associated with the collection and analysis of representative samples of dredging slurries have been discussed by Raphaelian and Krizek.[2]

TEST PROGRAM

Tests were conducted on these samples to determine their (a) clay mineralogy, (b) calcium carbonate content with percentage of calcite and dolomite, (c) pH, (d) organic carbon content, (e) cation exchange capacity, (f) relative concentrations of seven ions, and (g) concentrations of 12 pesticides in both the fluid and solid portions of the samples.

X-Ray diffraction analyses were used to identify the clay minerals present. The relative amounts of each mineral in the -2μ fraction are based on integrated peak intensities and mass absorption coefficients, and the percentages of clay mineral represent rough estimates of the fractions of the clay portion, which consists of various clay minerals. The Furman gravimetric method[3] was used to test for carbonates, and the Walkley-Black method[4] was employed in the organic carbon test. The cation exchange capacity was conducted according to ASTM, STP 241,[5] and it was recorded as milliequivalents per 100 grams of dry material. The relative amounts of exchangeable ions were determined by use of atomic absorption and flame emission from the extract of the material from the cation exchange capacity test; the ions are given as milliequivalents relative to calcium. The results of the foregoing tests are summarized in Table 9.1.

Table 9.1 Characterization Analyses

	Sample			
	C2DM	C3TO	C5CO	D4CO
Clay Mineral (%)				
Illite	60	50	60	75
Kaolinite	20	15	20	25
Mixed layer	20	35	20	Trace
Carbonates (%)	29.3	35.6	38.6	39.9
Calcite (%)	19.0	19.7	–	–
Dolomite (%)	10.3	15.9	–	–
Hydrogen ion concentration (pH)	7.6	7.4	8.0	7.8
Organic carbon (%)	8.8	3.6	10.5	8.8
Cation exchange capacity (meq/100 g of dry material)	15.1	18.2	15.3	11.9
Ions (meq relative to meq of calcium)				
Aluminum (Al)	3.3	1.7	2.8	2.6
Calcium (Ca)	1	1	1	1
Magnesium (Mg)	0.31	0.17	0.21	0.19
Iron (Fe)	0.008	–	–	–
Sodium (Na)	0.170	00.060	0.050	0.070
Potassium	0.030	0.018	0.030	0.028
Manganese (Mn)	Trace	Trace	Trace	Trace

In the chlorinated pesticide tests the samples were allowed to settle for several hours, after which as much water as possible was pipetted from each sample without removing any of the bottom sediments. This fluid was used for the analyses of the water phase. Then a portion of each of the bottom sediments was removed and the water was allowed to drain from the solids before being tested. These bottom sediments were placed on aluminum foil and allowed to air-dry at room temperature. The results of these pesticide tests are presented on an air-dry basis in Table 9.2.

DISCUSSION

At best, obtaining and analyzing representative samples of dredging slurries is an extremely complicated task, due largely to the high variability frequently observed in the chemical composition of these materials and the complex nature of the interactions and interferences that may

Table 9.2   Pesticide Analyses[a]

Pesticide (ng/kg)	Sample			
	C2DM	C3TO	C5CO	D4CO
Bottom Sediments				
Lindane	35,309	33,194	1,819	24,272
Heptachlor	70,955	9,563	24,818	45,048
Aldrin	44,876	32,566	28,959	74,584
Heptachlor epoxide	98,645	43,335	82,463	148,831
Methoxychlor	1	1	1	1
Dieldrin	1	1	1	1
Endrin	16,365	1	33,862	18,071
o, p-DDE	120,563	47,379	114,583	185,355
p, p'-DDE	110,053	43,127	97,424	196,931
o, p-DDD	68,148	20,964	111,111	111,760
p, p'-DDT	89,293	34,305	145,653	110,637
o, p-DDT	121,637	15,558	32,669	24,417
Supernatant Waters				
Lindane	76	5	41	559
Heptachlor	113	11	18	244
Aldrin	62	4	66	2,569
Heptachlor epoxide	464	24	201	1,180
Methoxychlor	1	1	1	1
Dieldrin	115	1	1	1
Endrin	1	1	18	1
o, p-DDE	288	23	313	1,109
p, p'-DDE	216	9	536	1,513
o, p-DDD	62	7	175	380
p, p'-DDT	100	11	110	412
o, p-DDT	82	5	84	436

[a]No separation of pesticides, PCB's or phthalates was made. Therefore, these values may include the above.

occur. Hence, even if an overall phenomenon were delineated, the possible explanations for this behavior would be difficult, if not impossible, to verify. For example, bottom sediments may contain materials that act as molecular sieves; one such material is Chalazite, one of the naturally occurring zeolites. Alternatively, one might expect to find adsorbents such as alumina in the sediments, or ion exchange phenomena may take place. Based solely on phenomenological observations, one comprehensive study[6] indicated that the values of many pollution parameters (i.e., total solids, volatile suspended solids, BOD, COD, and several metals) measured for the supernatant waters were several orders of magnitude lower than those of the slurries, but apparently no studies of this type have been reported for pesticides.

Based on the test results given in Table 9.2, it can be seen that the concentrations of pesticides in the solid sediments are several orders of magnitude greater than corresponding concentrations in the supernatant waters from these sediments. Furthermore, the ratios between the various concentrations of particular pesticides in the solid portions of the samples to those in the liquid portions are greatest for Sample C3TO and smallest for Sample D4CO, with intermediate values for Samples C2DM and C5CO. Attempts to correlate this observation with the data given in Table 9.1 indicate that the pesticide content in the supernatant water decreases as (a) the illite content decreases, (b) the mixed layer clay content increases, and (c) the cation exchange capacity increases. However, there is no apparent correlation with organic carbon, probably because the pesticides comprise only a small fraction of the total organic contaminants in the sediments. Despite these observed associations, the specific nature of the mechanism or mechanisms (molecular sieve, adsorption, ion exchange, etc.) responsible for tying up the majority of the pesticides in the solids cannot be identified positively, and definitive correlations cannot be advanced from the available data.

CONCLUSIONS

Within the scope and limitations of the test program and associated interpretations presented herein, it can be reasonably concluded that the disposal of polluted bottom sediments within diked enclosures serves to contain the majority of the pesticides present. Pesticide concentrations in the supernatant or outflow waters are several orders of magnitude lower than those in the retained solid materials. Furthermore, it appears that bottom sediments with a high illite content, low mixed layer clay content and low cation exchange capacity tend to retain less of the pesticide content with the solids, but no positive explanation can be advanced for this phenomenon.

ACKNOWLEDGMENTS

This work was supported in large part by the Environmental Protection Agency under Grants 15070-GCK and R-800948. The pesticide analyses were obtained through the courtesy of Dr. Frederic D. Fuller, Illinois District Office of EPA, and the other characterization tests were performed by Dr. Gilbert L. Roderick, University of Wisconsin at Milwaukee, and other colleagues.

REFERENCES

1. Hummel, P. L. and R. J. Krizek. "Sampling of Maintenance Dredgings," *J. Test. Eval.* 2(3), 139 (1974).
2. Raphaelian, L. A. and R. J. Krizek. "Collection and Analysis of Representative Samples of Dredging Slurries," *Proc. Symposium on Water Resources Instrumentation*, International Water Resources Association, Vol. 1 (1974), p. 400.
3. Furman, N. H. *Standard Methods of Chemical Analysis.* Vol. 1, 5th ed. W. W. Scott, Ed. (New York: D. Van Nostrand Co., 1939).
4. Walkley, A. and I. A. Black. "An Examination of the Degtjareff Method for Determining Soil Organic Matter and a Proposed Modification of Chronic Acid Titration Methods," *Soil Sci.* 37, 29 (1934).
5. American Society for Testing and Materials. "Symposium on Exchange Phenomena in Soils," Special Technical Publication 142 (1952).
6. Krizek, R. J. and G. M. Karadi. "Disposal of Polluted Dredgings from the Great Lakes Area," *Proc. First World Congress on Water Resources*, International Water Resources Association, Vol. 4 (1973), p. 482.

10

A PRACTICAL SELECTIVE EXTRACTION PROCEDURE FOR SEDIMENT CHARACTERIZATION

Robert M. Engler, James M. Brannon, Jana Rose

U. S. Army Engineer Waterways Experiment Station
Environmental Effects Laboratory
Vicksburg, Mississippi 39180

Gary Bigham

Tetra Tech
Pasadena, California 91107

INTRODUCTION

Current interest in the possible alteration of water quality due to the natural resuspension of sediments or by man's relocation of sediments have warranted development of a realistic, reproducible, selective chemical extraction procedure for sediment characterization. The procedure would be a valuable research tool to assess the potential for alteration of water composition accompanying sediment resuspension.

Fundamental to understanding the impact of sediment resuspension on water quality is an understanding of how various elements that can have various effects on organisms, are associated with sediments.

Sediments can be divided into several components or phases that are classified by their composition and mode of transport to the estuarine environment. Among them are detrital and authigenic phases.

Detrital components are those that have been transported to a particular area, usually by water. Detrital materials are derived from soils of the surrounding watershed and can include (a) mineral grains and rock fragments (soils particles) as well as stable aggregates, (b) associated organic material, and (c) culturally contributed components derived from agricultural runoff and industrial and municipal waste discharged.

163

Authigenic components are those that formed in place or have not undergone appreciable transport. These materials are generally the result of aquatic organisms and include shell material ($CaCO_3$), diatom frustules (SiO_2), some organic compounds and products of anaerobic or aerobic changes.

In considering the *in situ* association with various sediment phases of trace elements in estuarine sediments, the water contained in inter-particle voids or interstices must be considered. This is termed inter-stitial water (IW). In relation to the overlying water, heavy metals are frequently enriched in the IW by several mechanisms. Heavy metals are relatively loosely bound to the sediment in several exchange locations; these include the exchange sites of the silicate phase and exchange sites associated with organic matter or trace elements complexed with the organic phase. Heavy metals are also associated with hydrated manganese, iron oxides and hydroxides, which are present in varying amounts in the sediment. Another location for heavy metals is in the sediment organic phase. The metals are incorporated into living terrestrial and aquatic organisms and are relatively stable but can be released into the sediment-water system during decomposition.

From the previous discussion of elemental partitioning and for analytical purposes, the following categories of sediment components will be considered.

1. *Interstitial water (IW).* This water, an integral part of sediment, is in dynamic equilibrium with the silicate and organic exchange phases of the sediment as well as with the easily decomposible organic phase.

2. *Mineral exchange phase.* That portion of the element that can be removed from the cation exchange sites of the sediment using a standard ion exchange extractant (NH_4OAc, dilute HCl, NaCl, $MgCl_2$, etc.).

3. *Reducible phase.* This phase is composed of hydrous oxides of iron and manganese as well as hydroxides of Fe and Mn, which are relatively stable under aerobic conditions but which come into solution under reducing (anaerobic) conditions. Of particular importance are the toxic metals (Zn, Cu, Cd, Ni, Co and Hg) that may be associated with these discrete Fe or Mn phases as occlusions or coprecipitates.

4. *Organic phase.* This phase or partition of elements is that considered to be solubilized after destruction of the organic matter. This phase contains very tightly bound elements as well as those loosely chelated by organic molecules. An initial extraction by an organic chelate may be needed to differentiate between the loosely bound and tightly bound elements.

5. *Residual phase.* This phase contains primary minerals as well as secondary weathered minerals which are, for the most part, a very stable portion of the elemental constituents. Only an extremely

harsh acid digestion or fusion will break down this phase. By far the largest concentration of metals is normally found in this fraction.

A particular element or molecule can be present (be partitioned) in a sediment in one or more of several locations. The possible locations include (a) the lattice of crystalline minerals, (b) the interlayer positions of phyllosilicate (clay) minerals, (c) adsorbed on mineral surfaces, (d) associated with hydrous iron and manganese oxides and hydroxides, which can exist as surface coatings or discrete particles, (e) absorbed or adsorbed with organic matter, which can exist as surface coatings or discrete particles, and (f) dissolved in the sediment interstitial water. These locations also represent a range in the degree by which an element may become dissolved in the receiving water. This range extends from stable components in the mineral lattices, which are essentially insoluble, to soluble compounds dissolved in the sediment interstitial water, which are readily mobile. Electrochemical (Eh, pH) changes after disturbing and resuspending anaerobic bottom sediments may result in possible solution or precipitation of many elemental species and should be thoroughly characterized.

To be practical, a procedure must be applicable to many types of marine and freshwater sediments, both aerobic and anaerobic. To be realistic, disturbance must be minimal. Thus, drying, grinding and contact with atmospheric oxygen are undesirable.

Such a technique has been developed and subjected to preliminary evaluation. Although we expect that minor modifications may be necessary, with the current interest this procedure will be useful to many investigators.

The effects of atmospheric oxidation on chemical transformations within sediments and interstitial water are well-known.[1-4] Previous sediment extraction studies[5-7] have not accounted for oxidation, which may cause phase or fraction differentiation of elements within the sediment.

Sediments may then be conceptually separated into several phases where chemical constituents can be extracted as a function of the analytical procedure. Quantitative knowledge of the selective distribution of chemical forms in the sediment aids in determining the relative availability of these constituents to biological communities and their availability to enter into reactions and transformations. These functionally defined phases are those dissolved in interstitial water, adsorbed (exchangeable) on sediment material, intimately associated with iron and manganese oxide and hydroxide phases (reducible), bound in organic matter and residual in a mineral crystalline lattice.

The exchangeable phase can be determined by several functionally derived ion exchange extraction procedures;[4,5,8-13] however, to inhibit phase differentiation due to oxidation and thus simulate *in situ* conditions, these extraction procedures must be conducted in an inert atmosphere.

The reducible (redox) fraction is that group of elements associated with and including the hydrous oxides of Mn and Fe, which come into solution under various physicochemical conditions.[4,5,9,10,13-19]

METHODS

Samples of fine-grained sediments are obtained by use of a gravity corer fitted with a plastic liner. This liner is sealed immediately after sampling to prevent oxidation. The cores are then stored upright at 4°C until returned to the laboratory for characterization. Figure 10.1 is a schematic representation of the sediment characterization procedure. Extrusion of the core and the extractions for interstitial water and exchangeable phases are made sequentially in a disposable glove bag in oxygen-free conditions verified with a polarographic oxygen analyzer.

Interstitial Water Phase

The sediment is extruded from the core liner into a plastic retainer and sectioned at desired depth intervals. Each section is then split into halves with each section receiving individual treatment. One half of a section is placed in an oxygen-free, polycarbonate, 500-ml centrifuge tube in the glove bag. Centrifugation in a refrigerated centrifuge at 9000 rpm for 5 minutes (13801 g's) at 4°C is sufficient to obtain approximately 40% of the total sediment water. The interstitial water so obtained is vacuum-filtered under nitrogen through a prerinsed 0.45μ membrane filter and immediately acidified to pH 1. The interstitial water is then stored in prerinsed plastic bottles for subsequent analysis.

Exchangeable Phase

The remaining half section is blended with an electrically driven polyethylene stirrer contained in the glove bag. Subsamples of the homogenized sediment section are weighed into oxygen-free, tared, 250-ml centrifuge tubes containing deoxygenated $1\,N$ ammonium acetate adjusted to the *in situ* sediment pH. The ratio of sediment to extractant is 1:5. Other subsamples of the homogenized sediment are removed for determination of percent solids. The centrifuge tubes are then shaken for one hour, and the exchangeable phase is separated by centrifugation and oxygenfreen vacuum filtration, then acidified and stored as described for the interstitial water. This phase also contains the interstitial water, and

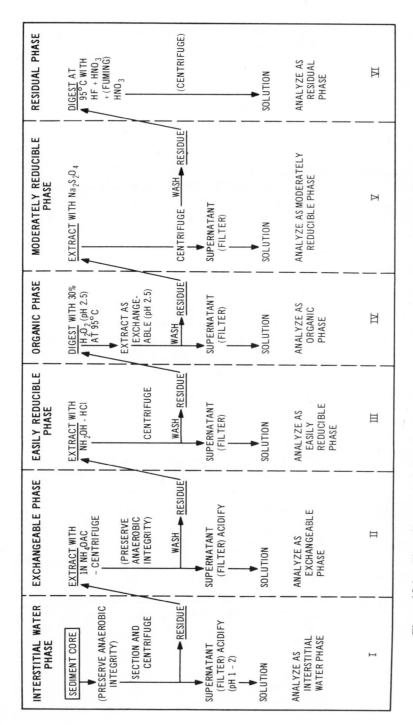

Figure 10.1 Flow diagram depicting the selective extraction procedure for chemical characterization of sediments.

specific concentrations of constituents must be corrected for concentrations found in the interstitial water.

Easily Reducible Phase

The residue of the exchangeable phase is washed once with deoxygenated water and centrifuged. Subsamples are taken from the washed, blended residue and extracted with acidified hydroxylamine hydrochloride for the hydrous manganese oxide phase and associated metals.[14] This extraction also removes a small (about 5%) fraction of the iron oxide system. The easily reducible fraction is separated by centrifugation and vacuum filtration. The residue is washed once with distilled deionized water and centrifuged. Metal carbonates and sulfides may also contribute to this phase.

Organic Phase

To break down the organic portion of the sediment, the residue from the previous portion is digested at 95°C with 30% H_2O_2 acidified to pH 2.5.[12] The digest is kept acidic to maintain solution of the metals. To remove the liberated ions from the mineral exchange sites, the acidified suspension is then extracted with $1 N$ ammonium acetate buffered at pH 2.5. The extract is separated as previously described, and the residue washed once with distilled water.

Moderately Reducible Phase

The washed residue from the preceding step is extracted with citrate-dithionite according to Holmgren.[15] The extractant reduces and extracts the hydrous iron oxides and associated metals that may be brought into solution, by using moderate-to-intense sediment-reducing conditions. The moderately reducible phase is separated by centrifugation and vacuum filtration. The residue is not washed with distilled water since this has been shown to extract additional iron.[20]

Residual

A subsample of the residue is digested over heat as described by Smith and Windom,[21] using HF, HNO_3 and HNO_3 (fuming). The solution is brought to volume in hot 1:1 HCl and is analyzed as the residual phase.

DISCUSSION

This selective dissolution scheme may be modified or refined to selectively extract carbonate minerals and metal sulfides (using acetic acid) as well as more refined fractions of the organic phase.

Percent moisture determinations should be made on sediment fraction subsamples prior to each extraction, with the exception of the moderately reducible phase, to determine the weight of solids involved in each step. The anaerobic integrity of the sample must be maintained throughout manipulation and extraction of the interstitial water, the exchangeable phase, and the first step of easily reducible phase.

The effects of temperature changes during sample collection and preparation are very important and should be considered.[22-26] If at all possible, the operational procedures should be conducted at *in situ* temperatures. If an *in situ* temperature cannot be maintained, storage and manipulation are best done at about 4°C. Kalil and Goldhaber[3] have proposed an inexpensive, portable sediment squeezer for operation at *in situ* temperatures, which precludes atmospheric oxidation.

The organic phase of the selective dissolution procedure was extracted prior to the extraction of the moderately reducible phase because an unknown portion of the organic fraction is dissolved by the citrate-dithionite. Complete oxidation of the sediment by the H_2O_2 treatment, however, does not appear to impair the efficiency of Holmgren's[5] method to reduce and extract the hydrous iron oxide fraction. Chao's[14] method for extraction of the easily reducible fraction did not appear to dissolve any organics; consequently, it was done prior to the H_2O_2 digestion.

Most so-called purified grades of sodium-dithionite are excessively contaminated with zinc; however, preliminary work has shown that some of the zinc can be removed using successive extractions of 3% ammonium pyrrolidine dithiocarbonate (APCD) in methyl isobutyl ketone (MIBK) without a decrease in the reducing activity of the dithionite. This solid-liquid extraction must be free of water, and the extracted dithionite should be well rinsed with APDC-free MIBK. Several hundred grams of sodium-dithionite may be purified in this manner. The lowest concentrations of sodium-citrate sodium-dithionite that will completely extract the iron should be used to minimize matrix interference during analysis for trace metals.

ACKNOWLEDGMENT

This research was funded under the Dredged Material Research Program of the U.S. Army Corps of Engineers.

REFERENCES

1. Connell, W. E. and W. H. Patrick. "Reduction of Sulfate to Sulfide in Waterlogged Soil," *Proc. Amer. Soc. Soil Sci.* 33, 711 (1969).
2. Hem, J. E. "Some Chemical Relationships among Sulfur Species and Dissolved Ferrous Iron," *Geol. Survey Water-Supply Paper* 1459-C (1960).
3. Kalil, E. K. and M. Goldhaber. "A Sediment Squeezer for Removal of Pore Waters without Air Contact," *J. Sed. Petrol.* 43, 553 (1973).
4. Patrick, W. H., Jr., S. Gotoh and B. G. Williams. "Strengite Dissolution in Flooded Soils and Sediments," *Science* 179, 564 (1973).
5. Chester, R. and M. G. Hughes. "A Chemical Technique for the Separation of Ferro-Manganese Minerals, Carbonate Minerals and Absorbed Trace Elements from Pelagic Sediments," *Chem. Geol.* 2, 249 (1967).
6. Nissenbaum, A. "Distribution of Several Metals in Chemical Fractions of a Sediment Core from the Sea of Okhotsk," *Israel J. Earth Sci.* 21, 143 (1972).
7. Presley, G. J., Y. Kolodny, A. Nissenbaum and D. R. Kaplan. "Early Diagenesis in a Reducing Fjord Saanich Inlet, British Columbia—II. Trace Element Distribution in Interstitial Water and Sediment," *Geochim. Cosmochim. Acta* 36, 1073 (1972).
8. Black, C. O. (Ed.). "Methods of Soil Analyses, Part 2," *Amer. Soc. Agron.* 9, Madison, Wisconsin (1965).
9. Gibbs, R. J. "Mechanisms of Trace Metal Transport in Rivers," *Science* 180, 71 (1973).
10. Gotoh, S. and W. H. Patrick, Jr. "Transformation of Manganese in a Waterlogged Soil as Affected by Redox Potential and pH," *Proc. Amer. Soc. Soil Sci.* 36, 738 (1972).
11. Gotoh, S. and W. H. Patrick, Jr. "Transformation of Iron in Waterlogged Soil as Influenced by Redox Potential and pH," *Proc. Amer. Soc. Soil Sci.* 38, 66 (1974).
12. Jackson, M. L. *Soil Chemical Analysis.* (Englewood Cliffs, New Jersey: Prentice-Hall, Inc., 1958).
13. Patrick, W. H., Jr. and F. T. Turner. "Effect of Redox Potential on Manganese Transformation in Waterlogged Soil," *Nature* 220, 476 (1968).
14. Chao, L. L. "Selective Dissolution of Manganese Oxides from Soils and Sediments with Acidified Hydroxylamine Hydrochloride," *Proc. Amer. Soc. Soil Sci.* 36, 764 (1972).
15. Holmgren, G. S. "A Rapid Citrate-Dithionite Extractable Iron Procedure," *Proc. Amer. Soc. Soil Sci.* 31, 210 (1967).
16. Jenne, E. O. "Controls on Mn, Ge, Co, Ni, Cu and Zn Concentrations in Water. The Significant Role of Hydrous Mn and Fe Oxides," in *Trace Inorganics in Water, Adv. in Chem.* 73, 337 (1968).
17. Krauskopf, K. B. "Factors Controlling the Concentrations of Thirteen Rare Metals in Sea Water," *Geochim. Cosmochim. Acta* 9, 1 (1956).
18. Loganathan, P. and R. G. Burau. "Sorption of Heavy Metal Ions by a Hydrous Manganese Oxide," *Geochim. Cosmochim. Acta* 37, 1277 (1973).

19. Mortimer, C. H. "Chemical Exchanges between Sediments and Water in the Great Lakes; Speculation on Probable Regulatory Mechanisms," *Limnol. Oceanog.* **16**, 387 (1971).
20. Brannon, J. M., J. R. Rose, R. M. Engler and I. Smith. "Distribution of Heavy Metals in Sediment Fractions from Mobile Bay, Alabama," Proc. Symp. on Chemistry of Marine Sediments, American Chemical Society, Atlantic City, New Jersey (1974).
21. Smith, R. G. and H. L. Windom. "Analytical Handbook for the Determination of As, Cd, Co, Cu, Fe, Pb, Mn, Ni, Hg and Zn in the Marine Environment," *Georgia Marine Science Center Technical Report 72-6* (1972).
22. Bischoff, J. L. and T. L. Sayles. "Pore Fluid and Mineralogical Studies of Recent Marine Sediment: Bauer Depression Region of East Pacific Rise," *J. Sed. Petrol.* **42**, 711 (1972).
23. Bischoff, J. L., R. E. Greer and A. O. Luistro. "Composition of Interstitial Waters of Marine Sediments: Temperature of Squeezing Effect," *Science* **167**, 1245 (1970).
24. Fanning, K. O. and M. E. Q. Pilson. "Interstitial Silica and pH in Marine Sediments: Some Effects of Sampling Procedures," *Science* **173**, 1228 (1971).
25. Mangelsdorf, P. C., T. R. S. Wilson and E. Daniell. "Potassium Enrichments in Interstitial Waters of Recent Marine Sediments," *Science* **165**, 171 (1969).
26. Scholkovitz, E. "Interstitial Water Chemistry at the Santa Barbara Basin Sediments," *Geochim. Cosmochim. Acta* **37**, 2043 (1973).
27. Friedman, G. M. and N. Kumar. "Procedure for Shipboard Measruements of pH and Eh in Sediment Cores within Plastic Liners," *J. Sed. Petrol.* **39**, 1247 (1969).
28. Howeler, R. H. "The Oxygen Status of Lake Sediments," *J. Environ. Quality* **1**, 366 (1972).

11

RELEASE OF TRACE CONSTITUENTS FROM SEDIMENTS
RESUSPENDED DURING DREDGING OPERATIONS

Thomas H. Wakeman*

Biological Oceanographer
U.S. Army Corps of Engineers District
211   Main Street
San Francisco, Californa   94105

INTRODUCTION

Sediments serve as a repository of organic and inorganic detritus which contain heavy metals.[1] These heavy metals can be original constituents of the detritus or can be incorporated during some transfer phenomena. Normally these metals are considered to be chemically stable in the sediments and unlikely to be released to the overlying waters.[2] However, a disturbance in environmental conditions may result in a shift in sediment-water equilibria, thereby affecting the mobilization of these heavy metals.

The dredging and disposal operation disturbs this sediment-water system, and thus could result in the release of metals or other toxicants.[3,4] Although much concern has been expressed, little has been done to determine if heavy metals are mobilized from the sediments during dredging or disposal operations. It has been assumed by some that they are and by others that they are not.

To determine whether the problem deserved further investigation within the localized conditions of San Francisco Bay, a survey experiment was performed. Its purpose was to qualitatively determine whether there were detectable increases in the heavy metal level in the water column during dredging or disposal operations. The experiment was conducted during the

---

*The views of the author do not purport to reflect the position of the Department of the Army or the Department of Defense.

1973 maintenance dredging of Mare Island Strait with disposal at
Carquinez Straits, San Francisco Bay (Figure 11.1). In the experiment,
water samples were collected in the dredged and disposal areas, with and
without the influence of the operations.

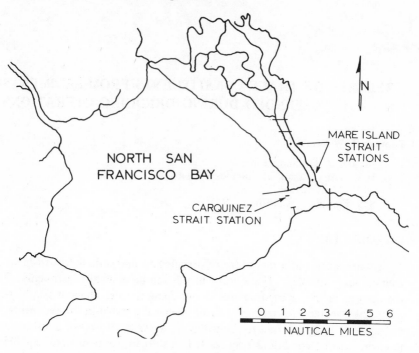

Figure 11.1 Location map.

Samples were collected on January 9 and 23, February 6 and 13, 1973,
using a three-liter PVC Van Dorn water sampler. Each sample was split
into two one-quart polyethylene containers, the first container to be analyzed
without filtering and the second container to be analyzed after filtering.
The unfiltered samples were acidified with 3 ml 1:1 HNO_3 per liter. The
samples were analyzed for chemical oxygen demand (COD), total Kjeldahl
nitrogen (TKN), mercury, zinc, lead, copper, chromium and nickel. The
mercury was analyzed on a Coleman Model 50 Hg Analyzer (cold vapor
technique) and the other metals on a Beckman Model 979 Atomic Absorption unit. Analyses were performed according to *Methods for Chemical
Analysis of Water and Wastes*,[5] at the Corps of Engineers South Pacific
Division Laboratory, Sausalito, California. Samples were taken at one
meter below the surface and one meter above the bottom on all dates
except January 9, 1973. On this day samples were taken only at the surface

because of sample container limitations. Three stations were sampled and they were: (a) Mare Island, 2135 meters south of the causeway, (b) Mare Island, 5180 meters south of the causeway, and (c) Carquinez Straight, on the south side of the disposal site. The January 23, 1973 samples were considered representative of background levels because they were taken following a 3-1/2 day period when the dredge was not working. This was assumed to be adequate time for any previous dredging effects to move out of the area or stabilize at the background levels.

Both COD and TKN increased during the operation. COD increases have been reported by Windom.[6] There are numerous papers on the phosphorus and nitrogen cycles in the sediment-water system. Ingle et al.,[7] Briggs,[8] and Cronin et al.[9] reported significant increases in nutrient concentrations adjacent to and downstream from dredging activities. Total phosphorus increases as high as 1000 times over background were recorded, and total nitrogen increases of 60 times. May[10] reported slight total phosphorus increases when water content was low and no increase when phosphorus levels in the water were high. Laboratory research[11,12] indicates the process is apparently concentration-dependent and that dissolved phosphorus is at equilibrium with oxidized sediments at approximately 95 μg/l.

The heavy metal results obtained from the sampling program were treated statistically. The mean and standard deviation of each parameter were calculated for the background period (January 23) and the study period (January 9, February 6 and 13). Because of the uniformity of the water in the channel, the data from the two Mare Island stations were combined. The reduced data are presented in Table 11.1.

The means of the background samples were compared to the means of the samples obtained during the operations using Student's "t" Test at the P = 0.05 level. Significant fluctuations from background are presented in Table 11.2.

The data show that the levels of zinc, lead, chromium and nickel were significantly greater during the period of dredging and disposal operations. Mercury decreased in one case and in all other cases showed no significant change. Copper levels did not change in any case.

Elutriate analyses of Mare Island Strait sediments were performed using the methods described by Keeley and Engler.[13] The results of the analyses are shown in Table 11.3. The mercury decreased in the laboratory test as it did in the field. Lead, and to a lesser extent copper, increased, which is also reflected in the field results. Cadmium was not analyzed in the field, but elutriate analysis showed an increase by a factor of two.

Table 11.1   Number of Samples, Mean and Standard Deviation of Each Parameter

Parameter		Hg	Zn	Pb	Cu	Cr	Ni
Background							
TUF[a]	n	3	3	3	3	3	3
	mean	0.23	0.17	0.08	0.11	0.04	0.04
	std. dev.	0.06	0.05	0.0	0.01	0.0	0.0
BUF[b]	n	3	3	3	3	3	3
	mean	0.2	0.17	0.08	0.11	0.04	0.04
	std. dev.	0.0	0.09	0.0	0.02	0.0	0.0
TF[c]	n	3	3	3	3	3	3
	mean	0.13	0.136	0.08	0.09	0.04	0.04
	std. dev.	0.05	0.005	0.0	0.04	0.0	0.0
BF[d]	n	3	3	3	3	3	3
	mean	0.1	0.1	0.08	0.07	0.04	0.04
	std. dev.	0.0	0.005	0.0	0.02	0.0	0.0
Mare Island Straits							
TUF	n	11	11	11	11	11	11
	mean	0.2	0.69	0.11	0.12	0.13	0.08
	std. dev.	0.13	0.29	0.03	0.05	0.05	0.04
BUF	n	5	5	5	5	5	5
	mean	0.2	0.58	0.13	0.09	0.12	0.15
	std. dev.	0.07	0.35	0.05	0.03	0.02	0.04
TF	n	11	11	11	11	11	11
	mean	0.12	0.29	0.09	0.08	0.07	0.05
	std. dev.	0.06	0.09	0.02	0.03	0.03	0.02
BF	n	5	5	5	5	5	5
	mean	0.16	0.26	0.08	0.08	0.09	0.12
	std. dev.	0.05	0.18	0.01	0.03	0.03	0.04
Carquinez Straits							
TUF	n	6	6	6	6	6	6
	mean	0.16	0.76	0.10	0.11	0.10	0.09
	std. dev.	0.05	0.24	0.02	0.05	0.05	0.19
BUF	n	3	3	3	3	3	3
	mean	0.2	0.73	0.14	0.18	0.14	0.18
	std. dev.	0.0	0.48	0.01	0.05	0.04	0.03
TF	n	6	6	6	6	6	6
	mean	0.13	0.26	0.09	0.08	0.06	0.08
	std. dev.	0.05	0.11	0.02	0.02	0.02	0.05
BF	n	3	3	3	3	3	3
	mean	0.1	0.25	0.13	0.08	0.08	0.16
	std. dev.	0.0	0.10	0.01	0.03	0.03	0.06

[a]TUF = top unfiltered sample.
[b]BUF = bottom unfiltered sample.
[c]TF = top filtered sample.
[d]BF = bottom filtered sample.

Table 11.2   Significant Increases (I) or Decreases (D) from Background Levels

Parameter	Hg	Zn	Pb	Cu	Cr	Ni
Mare Island						
TUF	−	I	I	−	I	I
BUF	−	−	−	−	I	I
TF	−	I	I	−	I	I
BF	−	−	−	−	I	I
Carquinez Straits						
TUF	D	I	I	−	I	−
BUF	−	−	I	−	I	I
TF	−	I	−	−	I	−
BF	−	−	I	−	−	I

Table 11.3   Elutriate Analyses of Mare Island Sediment

	Disposal Site	Elutriate Results No. Samples	$\bar{x}$	$S_{\bar{x}}$
Lead (mg/l)	0.012	19	0.020	0.013
Copper (mg/l)	0.020	19	0.021	0.005
Mercury, (mg/l)	0.0007	19	0.0001	0.0001
Cadmium (mg/l)	0.001	19	0.0020	0.0014

Several mechanisms have been suggested for control of heavy metal partitioning. Some of the more important mechanisms are redox,[14] iron and manganese oxide coatings,[15] salinity,[16] sulfides,[17] iron cycling,[6] organic matter[1] and pH.[2]

Presumptively, none of these mechanisms is solely responsible for the control of metal mobilization. When mobilization does occur, it is probably the result of a combination of these mechanisms that affect a different response from each metal. The absence of uniform metal behavior is supported by this data.

Investigations by May[10] and Windom[6] indicate that dredging operations do not result in massive heavy metal desorption. It is agreed generally that the heavy metals are tightly bound to organic and mineral particulates, and "There is no wholesale release of metals . . ." during most dredging operations.[2] The exact reasons for the elevated levels of the four metals in the water column during this dredging are unknown. They could have resulted directly from the dredging operations, or from unusual sediment-water conditions, i.e., climatic associated aberrations. Climatic conditions during the study period were abnormal. Precipitation was more than 41 cm while normally it would be only about 22 cm for that time of the year. Salinities were recorded at less than 1 ppt in the surface waters and greater than 12 ppt in the bottom waters. pH readings taken at the time ranged from 7 at the surface to 10 at the bottom. This unusual amount of precipitation did result in bypassing from the Vallejo Municipal Sewage Plant and could have caused other unreported industrial discharges. The nature of these discharges was unknown; thus their chemical responses when resuspended by the dredging operation were unpredictable.

The dredged sediments from Mare Island Strait are anaerobic, i.e., in a negative redox state.[18] The heavy metal ions in this zone of negative redox potential are free to migrate through the pore water because of their increased solubility in the reduced state.[6] During the resuspension of these anaerobic sediments, metals that are normally insoluble in an oxidized state could be released. Once released the metals would immediately adsorb onto organic matter, iron or manganese oxides. However, the environmental conditions in this case could have inhibited metal relocation.

This survey experiment was just one brief glimpse at the sediment-water system during the dredging and disposal operation. From it alone conclusions cannot be drawn. As a result of the experiment, the Corps of Engineers, San Francisco District, has performed a quantitative investigation of the sorption-desorption phenomenon[19] to identify the mechanisms and kinetics of heavy metal mobilization from resuspended Bay sediments.

REFERENCES

1. De Grott, A. J. and E. Allersma. "Field Observations on the Transport of Heavy Metals in Sediments," *Proc. Conf. on Heavy Metals in the Aquatic Environment*, Nashville, Tennessee (1973).
2. Lee, G. F. and R. H. Plumb. *Literature Review on Research Study for the Development of Dredge Material Disposal Criteria*, prepared for U.S. Army Corps of Engineers, Contract No. DACW39-74-C-0024, Vicksburg, Mississippi (1974), 130 pp.

3. Sherk, J. A. "Current Status of the Knowledge of the Biological Effects of Suspended and Deposited Sediments in Chesapeake Bay," *Chesapeake Sci.* **13**, S137 (1972).

4. O'Neal, G. O. and J. Sceva. *The Effects of Dredging on Water Quality in the Pacific Northwest*, Environmental Protection Agency, Region X (1971).

5. Environmental Protection Agency. *Methods of Chemical Analysis of Water and Wastes* (Cincinnati, Ohio: Water Quality Office, EPA, 1971), 312 pp.

6. Windom, H. L. "Environmental Aspects of Dredging in Estuaries," *J. Amer. Soc. Civil Eng.* **98**(WW4), 475 (1972).

7. Ingle, R. M., A. R. Ceurvels and R. Leinecker. *Chemical and Biological Studies of the Muds of Mobile Bay*, Report to Division Seafoods, Alabama Department of Conservation, University of Miami Contribution 139 (1955), 14 pp.

8. Briggs, R. B. "Environmental Effects of Overboard Spoil Disposal," *J. San. Eng. Div. ASCE* **94**(SA3), 477 (1968).

9. Cronin, L. E., R. B. Biggs, D. A. Flemer, H. T. Pfitzenmeyer, F. Goodwin, Jr., W. L. Dovel and D. E. Ritchie, Jr. *Gross Physical and Biological Effects of Overboard Spoil Disposal in Upper Chesapeake Bay*, Natural Resources Institute, University of Maryland, Contribution 397, Special Report 3 (1970), 66 pp.

10. May, E. B. "Environmental Effects of Hydraulic Dredging in Estuaries," *Alabama Mar. Res. Bull.* **9**, 1 (1973).

11. Rochford, D. J. "Studies in Australian Estuarine Hydrology. I. Introductory and Comparative Features," *Austral. J. Mar. Freshwater Res.* **2**, 1 (1951).

12. Pomeroy, L. R., E. E. Smith and C. M. Grant. "The Exchange of Phosphate between Estuarine Water and Sediments," *Limnol. Oceanog.* **10**(2), 167 (1965).

13. Keeley, J. W. and R. M. Engler. "Discussion of Regulatory Criteria for Ocean Disposal of Dredged Materials: Elutriate Test Rationale and Implementation Guidelines," Miscell. Paper D-74-14, U.S. Army Corps of Engineers Waterways Experiment Station (1974) 13 pp.

14. Brooks, R. R., B. J. Presley and I. R. Kaplen. "Trace Elements in the Interstitial Waters of Marine Sediments," *Geochim. Cosmochim. Acta* **32**, 397 (1968).

15. Jenne, E. A. "Controls on Mn, Fe, Co, Ni, Cu, and Zn Concentrations in Soils and Water: The Significant Role of Hydrous Mn and Fe Oxides," In *Trace Inorganics in Water*, Amer. Chem. Soc. Adv. Chem. Series 73, ,Washington, D.C. (1968), p. 337.

16. Kharker, D. P., K. K. Turekian and K. K. Bertine. "Stream Supply of Dissolved Silver, Molybdenum, Antimony, Selenium, Chromium, Cobalt, Rubidium and Desium to the Ocean," *Geochim. Cosmochim. Acta* **32**, 285 (1968).

17. JBF Scientific Corporation. *Interaction of Heavy Metals with Sulfur Compounds in Aquatic Sediments and in Dredged Material*, prepared for U.S. Army Corps of Engineers, Contract No. DACW33-74-M-0190, Waltham, Massachusetts (1973), 14 pp.

18.   Brown and Caldwell Engineers. *Effects of Dredged Materials on Dissolved Oxygen in Receiving Water*, prepared for U.S. Army Corps of Engineers, Contract No. DACW07-73-C-0051, San Francisco (1973), 24 pp.

19.   Serne, R. J. and B. W. Mercer. *Characterization of San Francisco Bay Dredged Sediments—Crystalline Matrix Study*, prepared for U.S. Army Corps of Engineers, Contract No. DACWO7-73-C-0080, San Francisco (1975), 216 pp.

12

MIGRATION OF CHEMICAL CONSTITUENTS IN
SEDIMENT-SEAWATER INTERFACES

James C. S. Lu and Kenneth Y. Chen

Environmental Engineering Program
University of Southern California
Los Angeles, California 90007

INTRODUCTION

Marine sediments are known to be the sink of most suspended solids in the water column.[1,2] The function of sediment to act as a reservoir for nutrients has been extensively reported.[3-6] In addition, sediment has been proved to be a very effective sink for chlorinated hydrocarbons.[7-9] The role of sediments related to the migration of chemical components from sediment to seawater under disturbance is less conclusive.

The transport of trace metals from sediment to seawater may be a result of diffusion, desorption, chemical reaction, or biological effects. Under reducing conditions, Mn may diffuse to the overlying seawater from sediment.[10] In conducting leaching tests, Johnson *et al.*[11] found that about one-third of ^{54}Mn was released to the seawater. Data collected by Li and co-workers[12] also show that after burial, Mn may be remobilized under local reducing conditions. Other studies[13-15] show enrichment of Mn and other trace metals such as Cd, Co, Cu, Fe, Ni and Zn in the interstitial waters. Lee and Plumb[16] reviewed the literature to predict the release of trace contaminants and nutrients, and concluded that most trace metals could potentially be mobilized with the disturbance of sediment. Wakeman[17] found that Cr, Ni, Pb and Zn were significantly released during the period of a dredging and disposal operation.

The release of nutrients can be effected from the biodegradation of organic matter deposited on the sediment. Waksman and Hotchkiss[18] showed that the organic matter in the marine sediment can undergo slow but gradual oxidation due to biological activities. Mortimer[3,4] suggested that the oxidation-reduction condition can affect the release rate of nutrients from lake mud. Rittenberg et al.[19] found that ammonia can be regenerated from sediments and oxidized to nitrate if the Eh value in the overlying water is positive, while phosphate can be regenerated under anaerobic conditions. Kemp and Mudrochova[20] found that a minimum of 20% of the organic nitrogen can be regenerated and released to the overlying lake water from the top 6 cm of sediments. Austin and Lee[21] found that the concentration of NH_3-N increased under anaerobic conditions, but will remain at low levels under aerobic conditions. NO_3^- was found to be increased to a much higher level in the aerobic system.

In studying the release of radioactive phosphorus, Zicker[22] found that the amount of phosphorus released to the water was very small, with virtually no release from depths greater than 1/4 in. below the mud surface. MacPherson et al.[23] reported that the release of inorganic-P is pH-dependent. It was found that between pH 5.5 and 6.5, phosphorus can be released at the level of 0.2 ppm. In more acid or alkaline solutions, soluble phosphorus is found to increase to the level of 0.3-0.5 ppm. These values are consonant with the experimental results of Pomeroy et al.[24] in that the exchange in the sediment-water interface tends to maintain a concentration of about one μm phosphate/liter in the solution phase. McKee et al.[25] showed that the amount of phosphorus transported is proportional to the exposed surface area rather than to the quantity of the sediment. Li et al.[26] further indicated that the release of inorganic-P into solution under anaerobic conditions is mainly due to the reduction of Fe from ferric to ferrous state.

Similar to N or P, the Si species also may be released under appropriate conditions. Fanning and Schink[27] reported that sediments from the Atlantic sea floor release soluble silicate in large quantities when they are placed in contact with silica-poor seawater, but fail to reduce the concentration of enriched (211 μm) seawater. Elgawhary and Lindsay[28] concluded that there is a tendency for silicate to reach equilibrium value in the long run.

There is substantially less information available on the release of chlorinated hydrocarbons from natural sediments in comparison with studies on trace metals or nutrients. Under laboratory conditions, Dieldrin was found to be slightly released from sediment by washing with distilled water.[7,9] Nearly 70% of the adsorbed Lindane was leached in three successive washes with distilled water. The chlorinated pesticides were found to be tightly bound to sediments under most circumstances. It was concluded that the adsorption of chlorinated pesticides by clay minerals is accomplished predominantly by the formation of hydrogen bonding and other strong forces of interaction.[7]

Because of uncertainties of the controlling mechanisms, the transport phenomena in the sediment-seawater interface cannot be accurately described. Even though some qualitative information can be obtained from the literature, very little data have been obtained under well-defined environmental conditions.

In this study, efforts were made to control the redox conditions of the overlying waters to observe their effect on the migration of chemical constituents in the sediment-seawater interfaces.

METHODS

Sample Collection

Three types of sediments were collected from Los Angeles Harbor: silty clay, sandy silt and silty sand sediments. These sediment samples were taken with a box core. Upon extrusion, each sediment was sealed in plastic bags and stored in ice at 4°C for transport to the laboratory. The samples were then stored in a refrigeration unit at approximately 4°C until used.

Seawater was collected from a reference station about three miles outside the breakwater of the harbor (33°41.5′N, 118°14.5′W) in a polyethylene container. Water samples were usually used within 24 hours after collection.

The composition of sediments and the characteristics of the seawater are listed in Table 12.1 and Figures 12.1-12.18.

Experimental Setup

Two different types of long-term experimental tests were set up in a dark, constant-temperature chamber (10-14°C): first, disturbed sediment with resettling in a water column; and second, disturbed sediment without resettling (Figure 12.1).

Disturbed, Resettled Sediment Without Redox Control

Four liters of sediment were mixed with 16 liters of original seawater (without filtering) in a glove bag and shaken vigorously for about 10 minutes, then dumped into a 6-foot tall, 100-liter plexiglass cylinder with about 60 liters of seawater to make a final dilution ratio of 1:20 (v/v).

Table 12.1 Composition of Sediments
(All units in ppm unless specified)

Parameters[a]	Sandy Silt	Silty Sand	Silty Clay
TOC (%)	1.90	0.53	2.12
COD	52,290	29,210	116,800
IOD	538	383	1570
TVS (%)	4.59	2.80	10.1
$\Sigma S^=$	258	163	1670
Organic N	357	689	2820
Total N	357	706	2920
$\Sigma PO_4^{\equiv}$	886	679	1470
Ag	16.9	7.1	10.2
Cd	1.90	0.66	2.20
Cr	175	94	178
Cu	119	51.0	568
Fe	40,830	28,980	45,180
Hg	0.685	0.28	1.43
Mn	429	422	493
Ni	35.3	23.0	47.2
Pb	67	47	332
Zn	205	106	612
Particle size			
$< 5\ \mu m$ (%)	20	12	32
p,p'DDE	2.394	0.694	1.371
o,p'DDE	0.036	0.131	0.055
p,p'DDD	0.004	0.066	0.337
PCB 1254	0.252	0.110	0.366
PCB 1260	0.025	0.011	0.034
PCB 1242	0.192	0.121	0.304

[a]Other chlorinated hydrocarbons not listed in this table are below the detection limit.

After two days of resettling, the lower part of the column (20 liters, including most of the 4 liters of sediment) was removed. This reactor was sealed immediately, and placed in the constant-temperature environmental chamber for a long-term observation of the migration of chemical components in the interface.

Disturbed, Nonresettled Sediments with Redox Control

The same kind of 20-liter polyethylene reactor was used to observe the phenomenon of sediment-seawater interfaces without the presettling of sediments. Four liters of sediment were put into the reactor, then 16 liters

A. Flow Chart of Experiments

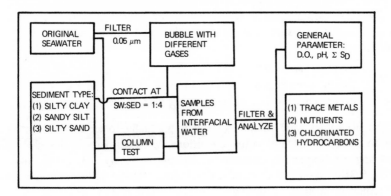

B. Long-Term Experimental Setup

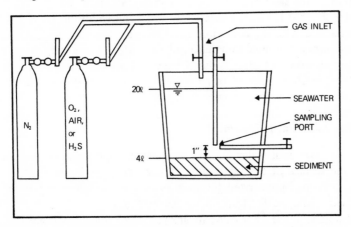

Figure 12.1   Flow chart of long-term experiments on the transport of trace metals between the sediment-seawater interfaces.

of purified and preconditioned seawater were carefully added along the wall of the reactor with a tygon tube. The purified and preconditioned seawater was prepared by passing the original seawater through 0.05 μm Millipore filter, followed by sparging with different ultrapurified gases (air, nitrogen or hydrogen sulfide) to render the seawater in an oxidizing, slightly oxidizing or reducing state. The ultrapurified gases were maintained in contact with the reactors (Figure 12.1) under the appropriate partial pressures throughout the experiment.

Experimental Conditions

The experimental conditions for the nonresettled test were checked at sampling times to insure that each system stayed within the desired conditions:

1. Oxidizing state: D.O. = 5-8 mg/l; ΣS_D = 0 mg/l
2. Slightly oxidizing state: D.O. = 0-1 mg/l; $\Sigma S_D \leqslant 0.05$ mg/l
3. Reducing state: D.O. = 0 mg/l; ΣS_D = 15-30 mg/l

The pH values were found to stabilize gradually to an equilibrium condition under different redox conditions. For oxidizing and slightly oxidizing conditions, pH decreased from approximately 8.3 to the range of 7-7.5 after about 15 days of contact time. Under reducing conditions, pH remained at about 7 during the experimental period.

For the resettled sediment without external controls, the environmental conditions of the seawater were changed from the original state. The interfacial water associated with the silty clay sediment was in an oxidizing condition from 0-1/4 day, in a slightly oxidizing condition from 1/4-3 days, and in a reducing condition after 3 days. For sandy silt and silty sand sediments, the conditions changed to a slightly oxidizing from an oxidizing environment after about 3 days of contact. In the sandy silt case, the condition changed to an oxidizing environment again (after about 45 days) due to a leakage in the lid of the reactor.

Sampling and Sample Treatment

All the interfacial water samples were taken approximately one inch above the surface of the sediment. To keep air from contaminating the sample, a syringe pressurized filtration technique and glove bag setup were used. Samples for trace metals analysis were passed through 0.05 μm membrane filter. For nutrient analysis, 0.45 μm membrane filter was used. For chlorinated hydrocarbons, glass filter was used.

Polyethylene bottles were used to store samples for trace metal and nutrient analysis, while pyrex glass bottles were used to store samples for the analysis of chlorinated hydrocarbons. Since the trace metals were at very low levels, special care was necessary to eliminate contamination. All containers for trace metal analysis were made of quartz, Teflon®,* polypropylene, or polyethylene materials and were cleaned by soaking in 5% acid for at least two days and multiple rinsing with high quality demineralized, distilled water.

The 0.05 μm membrane filter was found to be more satisfactory for the separation of soluble species from particulate matters. About 20 ml

*Registered trademark, E.I. duPont de Nemours & Company, Inc.

of filtrate was discarded before the filtrate samples were collected for analysis, to leach the membrane and eliminate possible water soluble contributions from the membrane filter.[29,30] To preserve the 0.05 μm membrane-filtered sample and break down complex forms, ultrapurified nitric acid was added to the sample to about pH 1. This acid-preserved sample was stored in a refrigerator and protected from light until time of analysis.

Nutrient samples were usually analyzed immediately after sampling. If prompt analysis was impossible, the nitrogenous species were acidified with sulfonic acid (100 μl in 100 ml sample) and stored at 4°C.[31] Samples for orthophosphate and total phosphorus were quick-frozen at 0°C,[32] and the dissolved silica samples were acidified with 0.3 ml of 1.0M HCl in 100 ml samples.[33]

Analytical Methods

pH, dissolved oxygen (DO), total dissolved sulfide (ΣS_D) and temperature were measured according to procedures described in *Standard Methods*.[34] Slight modification was made on the methylene blue photometric method for total dissolved sulfide, which was then checked by the sulfide electrode method.

The analytical methods for trace metals, nutrients and chlorinated hydrocarbons are listed as follows.

Trace Metals

A Perkin-Elmer atomic absorption spectrophotometer Model 305B with a heated graphite atomizer Model 2100 and deuterium arc background corrector was used. Trace metal analysis in seawater is obviously very complicated. In this study, a combination of methods,[35-39] together with some modifications, was used to obtain the most effective method of analysis. Following is a brief summary of the methods adopted:

	Method	Elements
1.	Direct injection	Cr, Fe, Mn
2.	APDC-MIBK extraction	Ag, Cd, Cu, Ni, Pb, Zn
3.	Flameless atomic absorption	Hg

Nutrients

Organic nitrogen (Kjeldahl method) and dissolved silica (heteropoly blue method) were analyzed according to *Standard Methods*.[34] Ammonia and nitrite were determined by the methods outlined by Riley *et al.*[33]

Nitrate was determined by the modified Brucine method.[40]   Orthophosphate and total phosphorus were determined by the amino-naphtholsulfonic acid method.[41]

Chlorinated Hydrocarbons

The extraction, separation and identification of chlorinated hydrocarbons were performed in accordance with the published literature.[42-49]   In all, about 20 species of chlorinated hydrocarbons have been analyzed: o,p'DDE, p,p'DDD, p,p'DDD, o,p'DDT, p,p'DDT; Arochlor 1242, 1252 and 1260; Lindane; BHC; Heptachlor; Aldrin; Heptachlor Epoxide; Kelthane; Methoxychlor; Chlordane; Toxaphene; Dieldrin; and Endrin.

RESULTS

The exchange phenomena between sediment-seawater interfaces differ widely under different environmental conditions.   Experimental data show that the direction of transport of trace metals from sediment to seawater or from seawater to sediment is controlled mainly by the redox conditions of the overlying seawater, and the flux is controlled chiefly by the type of sediment.   Nutrients were found to be released initially under all conditions.   As the contact time was increased, ammonia concentrations were found to be decreased under oxidizing conditions, while nitrate and nitrite disappeared under reducing conditions.   The concentrations of total phosphorus and orthophosphate in the interfacial water were found to decrease as contact time was increased.   Soluble silica was found to increase steadily.   In general, concentrations of trace metals in the interfacial waters were found to be at the sub-ppb to ppb levels, while the concentrations of nutrients were found to be in the sub-ppm and ppm ranges.   The amount of soluble chlorinated hydrocarbons was undetectable even after 3 months of contact.   The results of the long-term effects of redox conditions on the transport phenomena between different sediments are described as follows.

Transport of Trace Metals

Cadmium

Figure 12.2 shows that Cd can be significantly released only under oxidizing conditions, under which the Cd concentrations in the interfacial water were increased about 15 times over the original seawater background levels (about 0.03 ppb) to about 0.5 ppb after 4-5 months of contact.

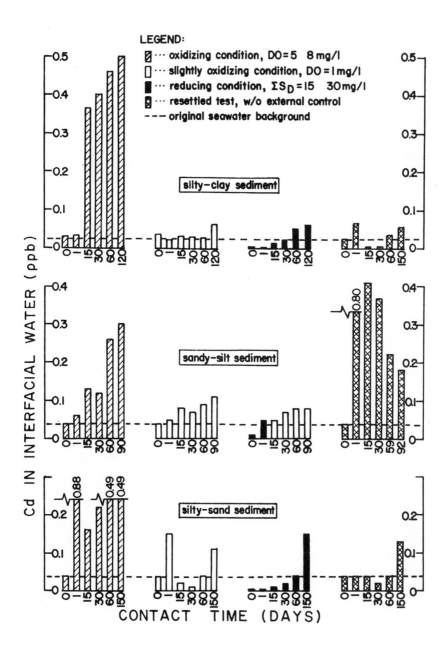

Figure 12.2 Transport of Cd between the sediment-seawater interfaces.

Under slightly oxidizing conditions, there was no significant change in Cd concentration. The concentration is close to the original seawater background. Under reducing conditions, the Cd concentration was at a very low level during the beginning contact period (about one week). But when contact time was increased, the concentration also increased.

No remarkable difference was noted between the different sediment types under identical environmental conditions. The silty sand sediment showed a spontaneous release followed by a deposition effect under both oxidizing and slightly oxidizing conditions. When contact time is increased, the trend of the exchange pattern tends to be the same.

Copper

All types of sediment seemed to remove Cu from the water in the beginning of the contact period (Figure 12.3). After one day of contact, the release phenomenon was observed under oxidizing and slightly oxidizing conditions. Under oxidizing conditions, Cu could be released to about 3 ppb from silty sand sediment after 5 months of contact, and to about 2 ppb from silty clay sediment after 4 months of contact. These concentrations are about 5-7 times higher than the original seawater background level (0.3-0.65 ppb). The original background Cu concentration for seawater in the sandy silt sediment experiment was very high (7.6 ppb); but after contact, a spontaneous scavenging effect reduced the Cu concentration to around the same original background level as found for the other two sediment types. When contact time increased, the Cu concentration again increased. The release rate of Cu from sediments took the following order:

silty sand > sandy silt > silty clay

Under slightly oxidizing conditions, the Cu concentration showed no significant difference from the original background. Under reducing conditions, the Cu concentration was decreased to an extremely low level. Redissolution of Cu then occurred, but at a very slow rate. For instance, after one month of contact in the silty sand test, Cu was increased to around the original background level. Under the same environmental conditions, the more clayey type of sediment is shown to release a lesser quantity of the Cu than other types of sediment, even though the Cu concentration in the clay sediment is the highest. For the same type of sediment, the higher the oxygen content in the interfacial water, the higher the release rate.

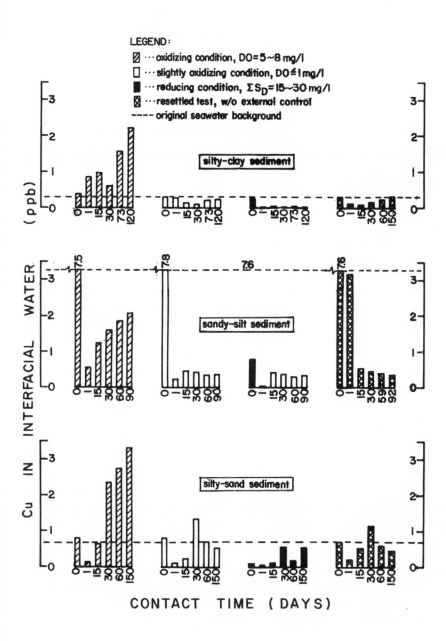

Figure 12.3   Transport of Cu between the sediment-seawater interfaces.

Chromium

The Cr concentration in the interfacial water shows no significant change under any redox condition in all of the sediment types.   The extremely slow rate of oxidation of Cr(III) to Cr(IV) in the aerated seawater environment explains the lack of redox effect.   Also, the absence of a chloride complex seems to limit any detectable release over the long term.

Iron

In general, Fe was found to be released substantially from all types of sediment under reducing conditions, and to be released slightly under oxidizing and slightly oxidizing conditions.   The order of the rate of release is

$$\text{reducing} > \text{slightly oxidizing} > \text{oxidizing}$$

Under oxidizing conditions, the final concentrations were about 2-5 ppb (Figure 12.4;   120, 90 and 150 days for the silty clay, sandy silt, and silty sand sediment, respectively).   But under slightly oxidizing conditions, the final concentrations could be raised to 10-15 ppb.   Under reducing conditions, the release rate of Fe was very rapid.   For instance, the silty clay sediment may increase the Fe concentration of the interfacial water from 0.5 ppb to about 100 ppb after 60 days of contact;   the silty sand sediment may increase the concentration even faster than the silty clay, from 0.2 ppb to about 800 ppb after 60 days.

Mercury

A comparison of Hg concentrations under different environmental conditions is shown in Figure 12.5.   In general, the Hg concentration showed little change from the original seawater background concentration.   Since the reported values were so close to the detection limit (0.02 ppb), the data seemed to be quite random.   However, the following pattern was still observed:

$$\text{oxidizing} > \text{slightly oxidizing} > \text{reducing}$$

The final concentration of Hg under oxidizing conditions was about 0.1 ppb.   Like Cd and Cu, under reducing conditions the Hg concentration decreased to a level below the detection limit.   The concentrations again increased after about 15 days of contact for silty sand sediment and after 2 months for the silty clay sediment.

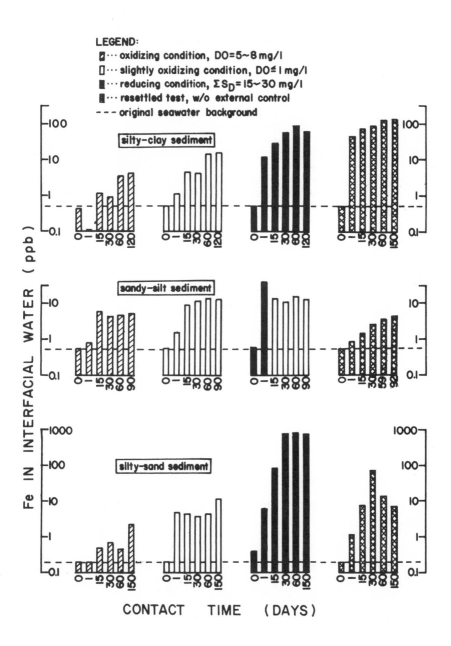

Figure 12.4 Transport of Fe between the sediment-seawater interfaces.

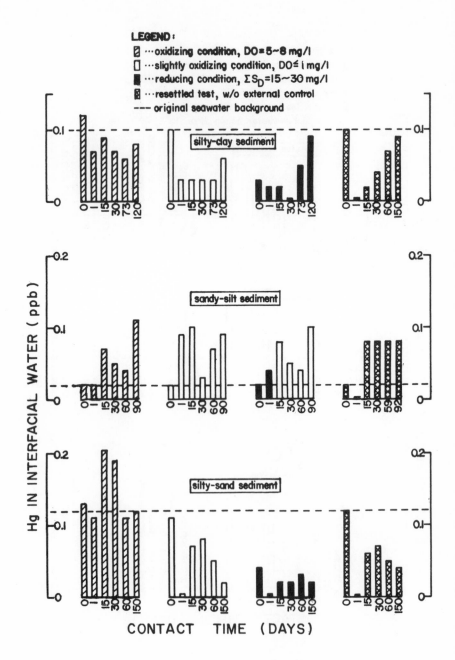

Figure 12.5 Transport of Hg between the sediment-seawater interfaces.

Manganese

The transport phenomena of Mn were similar to those of Fe, under different environmental conditions, which followed the release order

reducing $>$ slightly oxidizing $>$ oxidizing

(Figure 12.6). Under the same environmental conditions, the clayey-type sediment seemed to release less than the other sediment types. The release of Mn under reducing conditions was quite significant. For silty sand sediment in the final concentration, the Mn concentration was about 310 times that of the original background (0.25 ppb). For silty clay sediment, the concentration was about 85 times that of the original 0.3 ppb. Under oxidizing conditions, the final concentrations of Mn were less than 1 ppb in the silty clay and silty sand test, and about 7 ppb in the sandy silt test.

Nickel

In general, the concentration of Ni in the interfacial water took the order

oxidizing $>$ slightly oxidizing $>$ reducing

(Figure 12.7). If enough contact time were allowed, the Ni concentration seemed to be higher under reducing conditions than under slightly oxidizing conditions, as in the silty sand test. In this experiment, the sandy-type sediment could release more Ni. Under oxidizing conditions, the final Ni concentration was about 6 ppb for the sandy type of sediment and only about 2 ppb for the clayey type of sediment. Like Cu, in the reducing environment Ni also showed deposition during a short contact period. The concentration increased again when contact time was increased. In the clayey type of sediment, the Ni can be scavenged to undetectable levels during the first week of tests; the Ni will then be gradually redissolved. The redissolution concentration was quite high in the sandy sediments.

Lead

The transport phenomena of Pb followed the same trend as Cd, showing higher concentrations under oxidizing conditions (Figure 12.8). No significant change was observed under slightly oxidizing conditions, and a large decrease followed by a slight increase occurred under reducing conditions. Among these three conditions the concentration range is quite small, and the values never exceed sub-ppb levels even after five months of contact.

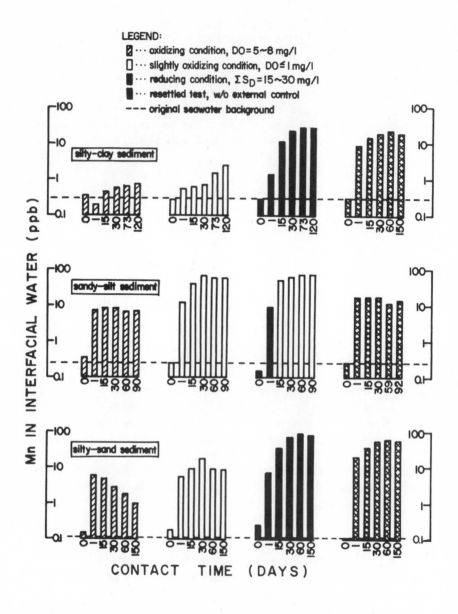

Figure 12.6 Transport of Mn between the sediment-seawater interfaces.

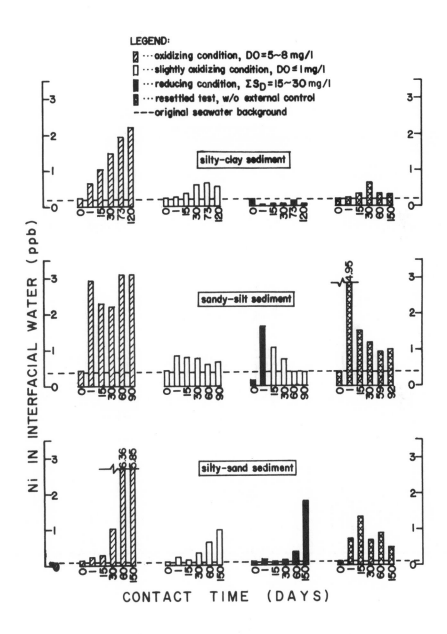

Figure 12.7   Transport of Ni between the sediment-seawater interfaces.

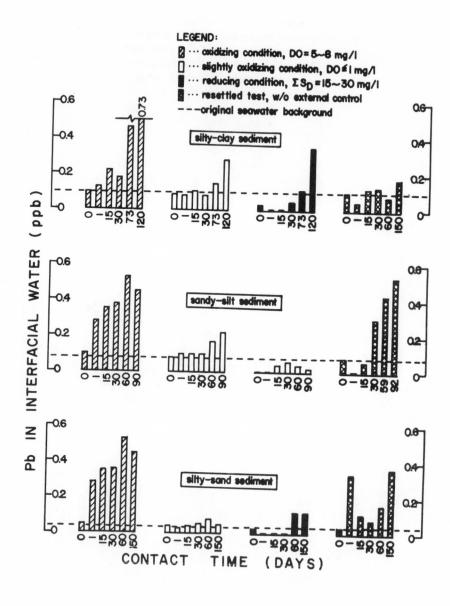

Figure 12.8   Transport of Pb between the sediment-seawater interfaces.

Zinc

Figure 12.9 shows that a fast release of Zn occurs after contact under both oxidizing and slightly oxidizing conditions. The clayey-type of sediment could release more Zn than the sandy-type sediment. Under an oxidizing environment, the final Zn concentration was about 10 ppb for the clayey-type sediment and only about 3 ppb for the sandy-type sediment. For silty sediment, the final Zn concentration was about 5 ppb. Under slightly oxidizing conditions, the final Zn level was about 2 ppb in all types of sediments. Under reducing conditions, the Zn concentration decreased to a very low level at the beginning of the contàct period and then increased again. Under a reducing environment, the redissolution phenomenon for the silty clay sediment test was very clear. The concentration of Zn was even higher than that of the original seawater background after about 70 days of contact.

Silver

The amount of Ag transported was undetectable (< 0.02 ppb) under all conditions and in all types of sediment during the experimental period.

Transport of Nutrients

Ammonia

Under different redox conditions, ammonia was found to be released in the following order:

$$\text{reducing} > \text{slightly oxidizing} > \text{oxidizing}$$

(Figures 12.10 to 12.12). Among the different sediment types, ammonia was found to be released in the following order:

$$\text{silty clay} > \text{sandy silt} > \text{silty sand}$$

Under oxidizing conditions, the concentration of ammonia reached 0.7 mg/l for silty sand and sandy silt sediment, and about 4.5 mg/l for silty clay sediment, followed by a decrease and eventual disappearance after a few months of contact. Under reducing conditions, NH_3-N was found to be continually released to about $1 \sim 2$ ppm for the silty and sandy sediments and to about 12 ppm for the clayey type of sediment.

Nitrite

Nitrite was present in the presence of oxygen (Figures 12.13 to 12.15). With the exception of clayey-type sediment, the NO_2-N was present in

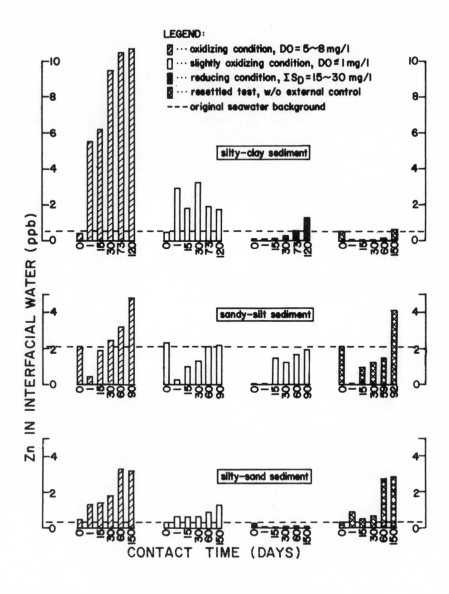

Figure 12.9   Transport of Zn between the sediment-seawater interfaces.

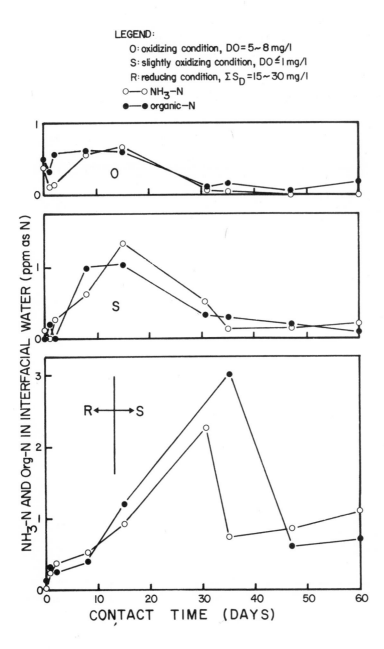

Figure 12.10 Transport of ammonia and organic nitrogen between the sediment-seawater interfaces (sandy silt sediment).

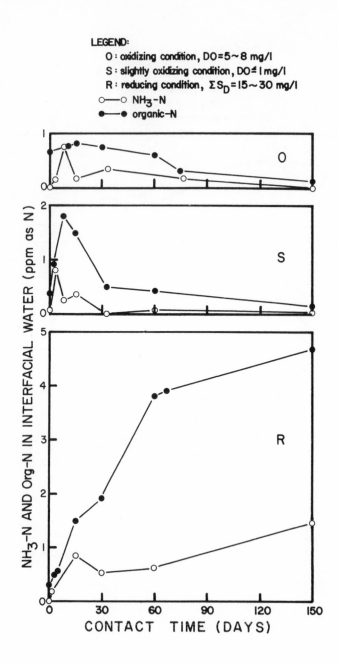

Figure 12.11   Transport of ammonia and organic nitrogen between the sediment-seawater interfaces (silty sand sediment).

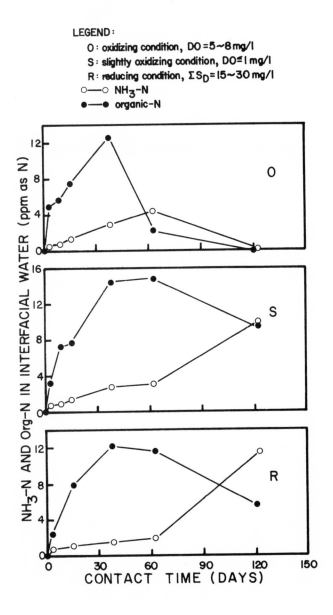

Figure 12.12 Transport of ammonia and organic nitrogen between the sediment-seawater interfaces (silty clay sediment).

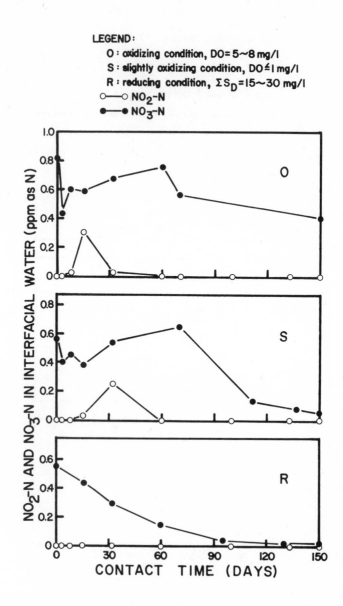

Figure 12.13   Transport of nitrate and nitrite between the sediment-seawater interfaces (silty sand sediment).

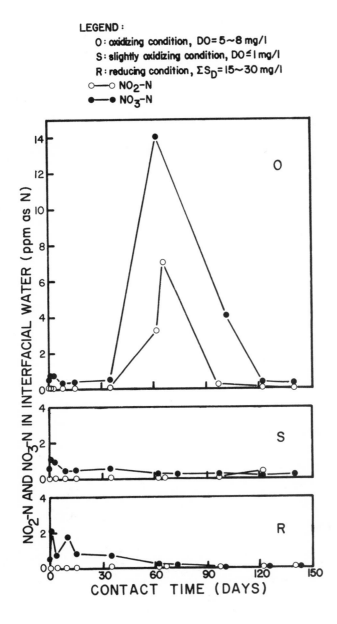

LEGEND :
O : oxidizing condition, DO = 5~8 mg/l
S : slightly oxidizing condition, DO ≤ I mg/l
R : reducing condition, ΣS_D = 15~30 mg/l
○——○ NO_2-N
●——● NO_3-N

Figure 12.14 Transport of nitrate and nitrite between the sediment-seawater interfaces (silty clay sediment).

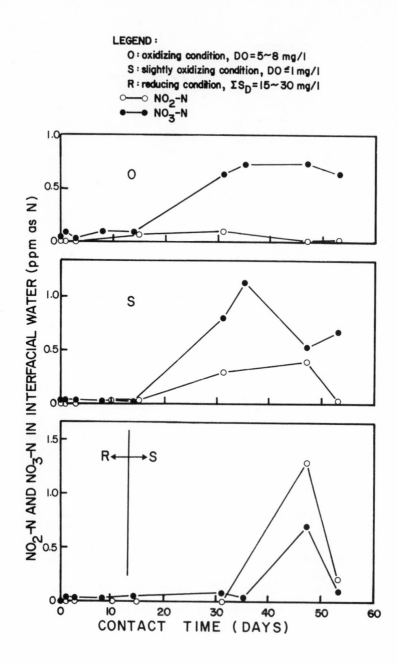

Figure 12.15 Transport of nitrate and nitrite between the sediment-seawater interfaces (sandy silt sediment).

the ppb range. Under reducing conditions, nitrite was undetectable throughout the experimental period.

Nitrate

Nitrate was present in the interfacial water in the presence of oxygen (Figures 12.13 to 12.15). Under oxidizing conditions, the NO_3-N concentration in the silty sand sediment was about 0.5 mg/l; but in the silty clay sediment test, the NO_3-N concentration reached as high as 14 mg/l. Obviously, the concentration in sediment phase determines the rate of release to a great extent. However, after about two months of contact, the nitrate concentration decreased again. Under reducing conditions, the nitrate concentration was found to decrease constantly, until it reached an undetectable amount.

Organic Nitrogen

For both silty sand and sandy silt sediments, the release of dissolved organic nitrogen was found to be higher under a reducing environment (Figures 12.10 to 12.12). Under oxidizing or slightly oxidizing conditions, the organic nitrogen content in the interfacial water of both sediments was found to decrease gradually to a trace amount. For silty clay sediment under all types of environments, the organic nitrogen could be released to about 10-15 mg/l, or about three times that of the reducing environment in the silty sand test. After about one month of contact time, the concentration of organic nitrogen in the silty clay test decreased again. Under oxidizing conditions, the organic nitrogen was found to be decreased to zero.

Total Phosphorus and Orthophosphate

Concentrations of both total phosphorus and orthophosphate in the interfacial water rose steadily during the beginning of the contact period under all environmental conditions (Figures 12.16 to 12.18). As contact time increased, there was a decreasing trend in both total phosphorus and ortho-PO_4. In general, no significant differences in release amounts were found among all types of sediments. But the environmental conditions of the overlying water did affect the exchange, and followed the order

$$\text{reducing} > \text{slightly oxidizing} > \text{oxidizing}$$

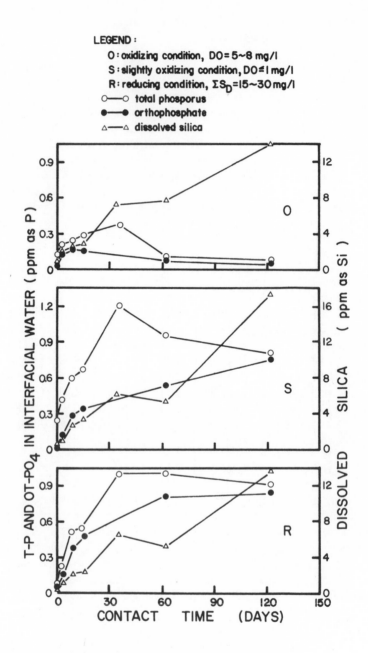

Figure 12.16 Transport of total phosphorus and orthophosphate between the sediment-seawater interfaces (silty clay sediment).

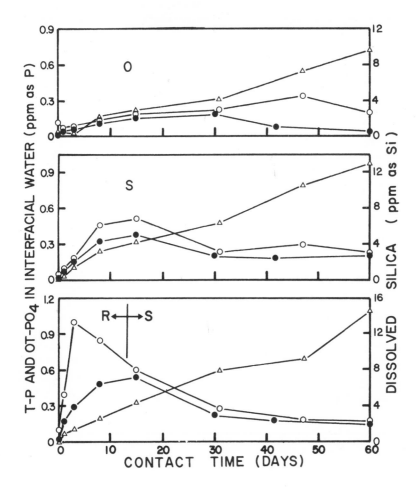

LEGEND:
O : oxidizing condition, DO = 5~8 mg/l
S : slightly oxidizing condition, DO ≦ I mg/l
R : reducing condition, Σ S_D = 15~30 mg/l
○——○ total phosporus
●——● orthophosphate
△——△ dissolved silica

Figure 12.17 Transport of total phosphorus and orthophosphate between
the sediment-seawater interfaces (sandy silt sediment).

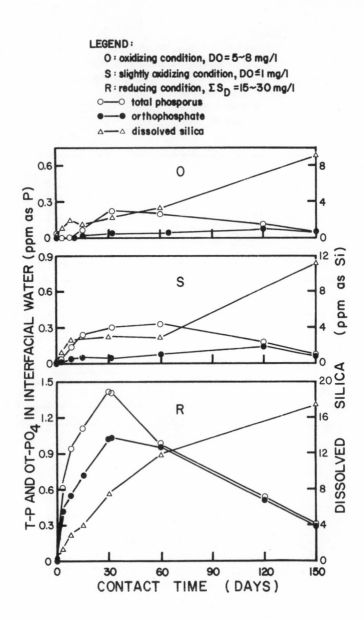

Figure 12.18   Transport of total phosphorus and orthophosphate between the sediment-seawater interfaces (silty sand sediment).

Dissolved Silica

The dissolved silica was found to be released continuously during the entire experimental period, eventually reaching the concentration of 8-16 mg/l. For the sandy and silty sediments, it seems that a reducing condition can release more dissolved silica. For silty clay sediment, no such distinction was observed (Figures 12.10 to 12.18). A comparison of the sediment types shows that the silty clay sediment has the highest release rate of dissolving silica, except under a reducing environment.

Transport of Chlorinated Hydrocarbons

The concentration of chlorinated hydrocarbons in the interfacial water of silty clay and silty sand sediments was analyzed over three months of contact time. The only species detected was PCB 1242, after eight days of contact time in the silty sand sediment test. The reason for this detection is unknown. The possibility of contamination is not ruled out.

Comparison of Experiments under Controlled and Uncontrolled Redox Conditions

There was no marked difference between these two types of experiments after redox effects are considered. The concentrations of trace metals and nutrients in the uncontrolled tests were close to those of similar redox conditions in the controlled tests. In the cases of silty sand and sandy silt sediments, the final environmental conditions of the uncontrolled test were close to the oxidizing condition of the controlled test. For silty clay sediment, it was close to the reducing condition.

DISCUSSION

Controlling Mechanisms for the Transport of Trace Metals

Oxidizing Conditions

Most of the trace metals, with the exception of Ag, Cr and Hg, were found to be released under oxidizing conditions. The possible mechanisms for the increase of soluble forms of trace metals in the interfacial water are

1. diffusion from interstitial water
2. desorption from clay minerals or other solid forms
3. chemical reaction, *i.e.,* oxidation of organics and sulfides
4. ion exchange

5. dissolution
6. complexation
7. biological actions

In these long-term experiments, physical and biological effects are considered only minor sources of soluble trace metals. The main release phenomenon is probably caused by complex formation. Under oxidizing conditions, through thermodynamic equilibrium calculations it can be shown that the concentration of soluble species of trace metals may be tremendously increased as their free ion forms are converted to complex forms.[50-52]

The main soluble forms of Cd may come from $CdCl_2$, $CdCl^+$, $CdCl_3^-$ and Cd^{+2}.[50] In addition, the presence of organic substances, hydroxide, fluoride and sulfate may form additional soluble complexes. This may explain the increase of Cd under oxidizing conditions.

The concentration of soluble mercury may also be controlled by chloride complexes ($HgCl_4^{-2}$, $HgCl_2$, and $HgCl_3^-$). But the release of Hg was not very significant in this experiment. It may be that the solubility of mercury was controlled by sulfides in the sediment. Despite the prevailing aerobic conditions in the water column, the diffusion of O_2 from the water column in this quiescent system may have been too slow or unable to oxidize the mercury solid species in the high-oxygen-demand sediments beneath the surface. Due to the relatively slow reaction between mercuric sulfide and oxygen, the mercuric sulfide may remain the predominant solid phase in the sediment.

Under oxidizing conditions, the soluble forms of Cu in the interfacial water may be mainly controlled by carbonate, borate and hydroxide.[58] For Pb, besides carbonate and hydroxide complexes, the chloride complexes also play an important role. For Zn, the main forms are probably Zn^{+2}, $Zn(OH)_2$ (aq), $ZnCl^+$, $ZnCl_2$ (aq), $ZnSO_4$ (aq) and organic complexes.[58] The Fe(II) and Mn(II) compounds are unstable under oxidizing conditions and oxidize to less soluble Fe(III) and Mn(IV) species. For this reason, the Fe and Mn concentrations show no significant increase in comparison with those under reducing conditions. The Cr concentration in the interface showed no significant change, probably as a result of the lack of soluble complexes and the low solubility of the hydroxide. A possible explanation for the undetectability of Ag is the extremely low solubility of AgS and AgCl. The most likely explanation for the lack of Ag release is the simple solubility consideration. Using a chloride concentration of 2% and a silver chloride product of 10^{-10}, it can be shown that the saturated silver concentration under these conditions would be slightly less than 0.02 $\mu g/l$, the detection limit of the instrument. If AgS is used as a controlling solid for Ag, the soluble Ag concentration is even lower.

The metals complexes that account for their high concentrations under oxidizing conditions are not solely those of chloride, hydroxide, carbonate or sulfate species. Others, such as organocomplexes, may also account for soluble forms in this experimental system. Given the complexity of the experimental system and considering that most complex formation constants are lacking, especially organometallic complexes, it is not possible to propose a more definitive model to explain the release phenomenon.

Reducing Conditions

Under a reducing environment, with the exceptions of Fe and Mn, most trace metals were decreased to an extremely low value at the beginning of the contact period. However, the concentrations of some of these metals (*e.g.*, Cd, Cu, Hg, Ni, Pb and Zn) increased again as time passed. It is suggested that the deposition effect during the beginning of the contact period is a combined result of metallic sulfide formation and adsorption, because the metallic sulfide species are more stable in comparison to most of the chloride, carbonate or other species. The subsequent increase of trace metals may be primarily due to the following:

1. formation of metallic sulfide complexes
2. formation of organometallic complexes
3. diffusion from interstitial water
4. release of trace metals from the dissolution of the highly insoluble hydrous metal oxides due to the formation of Fe(II) and Mn(II) compounds; and slow precipitation of metal sulfides due to kinetic effects.

Among these factors, the equilibrium of sulfide species may be the most important factor in determining the metal concentration in the interface for several species. On the other hand, organometallic complexes may play a role in cases where soluble concentrations exceed equilibrium calculations of metal sulfide complexes.

According to the stability constants from Ste-Marie *et al.*,[53] Barnes and Czamanske,[54] Schwarzenbach and Widmer,[55] Anderson,[56] and Sillen,[57] solubilization of metal species may be shown as follows, if the metallic sulfide is the controlling solid phase:

$$Ag_T = \left(\frac{K_{so}}{\alpha_2 \, \Sigma S_D}\right)^{1/2} [1 + K_2 K_1 \, (\alpha_1 \, \Sigma S_D)^2 + K_1 \, (\alpha_1 \, \Sigma S_D)]$$
$$+ \, 2K_{so2} \, (\alpha_1 \, \Sigma S_D)^2$$

$$Cd_T = \frac{K_{so}}{\alpha_2 \, \Sigma S_D} + K_{s10} \, (\alpha_0 \, \Sigma S_D) + K_{s11} \, \alpha_0 \, \alpha_1 \, (\Sigma S_D)^2$$
$$+ \, K_{s12} \, \alpha_0 \, \alpha_1 \, (\Sigma S_D)^2$$

$$Cu_T = \frac{K_{so}}{\alpha_2 \Sigma S_D} + K_{s12} \alpha_0 \alpha_1 (\Sigma S_D)^3$$

$$Hg_T = \frac{K_{so}}{\alpha_2 \Sigma S_D} + K_{s11} \alpha_0 \alpha_1 (\Sigma S_D)^2 + K_{so2} (\alpha_1 \Sigma S_D)^2$$
$$+ K_{s20} (\alpha_0 \Sigma S_D)^2 + K_{s1} (\alpha_2 \Sigma S_D)$$

$$Pb_T = \frac{K_{so}}{\alpha_2 \Sigma S_D} + K_{s11} \alpha_0 \alpha_1 (\Sigma S_D)^2$$

$$Zn_T = \frac{K_{so}}{\alpha_2 \Sigma S_D} + K_{s11} \alpha_0 \alpha_1 (\Sigma S_D)^2$$

where K_1 and K_2 are first and second ionization constants of H_2S, and

$$K_{so} = \{M^{Z+}\}^{\frac{2}{z}} \{S^{-2}\}$$

$$K_{sn} = \frac{\{M_{2/z} S_{n+1}^{-2}\}}{\{S^{-2}\}^n}$$
$$(n \neq 0)$$

$$K_{s1n} = \frac{\left\{M(HS)_{\frac{(n+2)z}{2}}^{-n}\right\}}{\{H_2S\}\{HS^-\}^n}$$
$$(n \neq 0)$$

$$K_{son} = \frac{\left\{M_{\frac{2}{z}} S(HS)_n^{-n}\right\}}{\{HS^-\}^n}$$

$$K_{sno} = \frac{\left\{M_{\frac{2}{z}} s(H_2S)_n\right\}}{\{H_2S\}^n}$$

where M^{Z+} = trace metals
$\quad\quad\quad$ Z = valence
$\quad\quad\quad$ n = positive integer

Under reducing conditions ($\Sigma S_D = 15 \sim 30$ mg/l; pH = $7 \sim 8$), only the concentrations of Cd, Hg and Pb were roughly close to the concentrations calculated by the above equations. The formation of sulfide complexes accounts for Cu concentrations of roughly 10^{-4} ppb, which is much lower than the actual figure (0.02 ~ 0.55 ppb). The subsequent increase of Cu in the interfacial water may be due to slow precipitation or nucleation, or may be due to the existence of organic complexes.

The concentration of Zn calculated from the above equation can again only account for about 10^{-4} of the actual concentrations. Obviously, the formations of soluble and organocomplexes account for most of the soluble zinc.[58]

Information on Ni-sulfide complexes is lacking. However, it is unlikely that the sulfide complexes account for the soluble concentrations observed.[58] The spontaneous release of Fe and Mn is obviously due to the existence of more soluble Fe(II) and Mn(II) compounds. It seems that FeS and MnS may be the most stable forms among inorganic ligands under sulfide-rich conditions. However, the results from calculations do not show such high solubilities. One possible reason may be that FeS and MnS colloids were so small that even 0.05 μm membrane filter could not capture them. The slow kinetics of precipitation or the presence of organocomplexes are two distinct possibilities.

Slightly Oxidizing Condition

In general, the concentrations of trace metals in interfacial seawaters under slightly oxidizing conditions were between the oxidizing and reducing conditions. Under oxygen-deficient conditions, due to the continued upward diffusion of dissolved sulfide from the sediment, the dissolved sulfide may overlap with the oxygen in the interface (as in silty clay sediment, where ΣS_D was approximately 0.05 mg/l). This dissolved sulfide may precipitate some of the dissolved trace metal species to metallic sulfides, and decrease the trace metal concentrations in soluble phase.

Controlling Mechanisms for the Transport of Nutrients

Oxidizing Condition

The results from Figures 12.10 to 12.12 show that under oxidizing conditions, organic nitrogen can be released to 1 ppm range for silty and sandy types of sediment, and up to 10 mg/l in clayey-type sediment. This seems to correlate well with the total concentration of organic nitrogen in the sediment phase. Further hydrolytic reactions will lead to the formation of ammonia and reduction in the concentration of organic nitrogen. The existence of ammonia will stimulate the autotrophic activities as follows:

$$NH_3\text{-}N \xrightarrow{\text{Nitrosomonas}} NO_2\text{-}N \xrightarrow{\text{Nitrobacter}} NO_3\text{-}N$$

Apparently, the diffusion of organic nitrogen from interstitial water, hydrolytic action on complex organic compounds and sorption phenomena all contribute to the concentration of organic nitrogen in the interfacial waters.

The NH_3-N is continuously converted to NO_2-N through biologically mediated oxidation. However, NO_2-N is short-lived in a strongly oxidizing environment. It will finally be converted to the only stable form, NO_3-N, in the oxidizing state. Finally, NO_3-N also seemed to be decreased, probably as a result of biological uptake. For these reasons, it is concluded that biological activity is the major force in regulating the concentration of nitrogen compounds in the sediment-water interfaces.

For the P-compounds, total phosphate may come mainly from the diffusion of interstitial water and hydrolytic reactions on complex organic compounds. Hydrolytic action on organic phosphate and condensed phosphate will increase concentration of orthophosphate under oxidizing conditions.

Diffusion from interstitial water and dissolution from solid forms toward equilibrium were the main driving forces for release of dissolved silica. Even after five months of contact, the equilibrium condition is apparently not reached.

Reducing Condition

The release mechanism for dissolved silica in the reducing condition is the same as that for the oxidizing condition. However, under reducing conditions $FePO_4$ solids will disappear, releasing Fe(II) and PO_4^{-3}. The route for nitrogen compounds under reducing conditions is reversed from that of oxidizing conditions through biologically mediated reduction. The organic nitrogen also converts to NH_3 through hydrolytic decomposition. NH_3 may also be derived from diffusion from interstitial water. As a result, in the experiments, NH_3-N concentration was continuously increased while other dissolved nitrogen compounds were decreased.

CONCLUSIONS

1. The experimental data show that the nature of sediment, *i.e.,* clay, silt or sand, does not control the direction of metal transport. It is regulated mainly by the chemistry of the immediately overlying water as well as that of the interstitial water. The redox chemistry is the principal factor. In general, the concentrations of trace metals in the interfacial water were found to be in the sub-ppb to ppb ranges.

2. Most nutrients were found to be released from the sediment to the overlying water under all conditions. The concentrations were found to be in the sub-ppm to ppm ranges.

3. There were no chlorinated hydrocarbons released from the sediments under any conditions, even after three months of contact time.

4. Under oxidizing conditions, with the exceptions of Ag, Cr, and Hg, all the observed trace metals were found to be released. The clayey-type sediment may release more Pb and Zn, while the sandy-type sediment may release more Cu and Ni. It is suggested that the release effect is primarily the result of chloride, hydroxide, carbonate and organocomplex formation.

5. Under oxidizing conditions, the nitrogen and phosphorus compounds increased during the beginning contact period (about one to two months) followed by a general decrease. NH_3 decreased to zero due to oxidation to NO_3^-. The concentration of silica was found to increase steadily. The clayey-type of sediment was found to release more nutrients. Diffusion, bio-oxidation, desorption and dissolution are the main regulating mechanisms.

6. Under reducing conditions, Fe and Mn were released at a very high level. The concentrations of other trace metals were decreased to extremely low levels in the initial contact period. As time passed, the concentrations of Cd, Cu, Hg, Ni, Pb and Zn were again increased. The deposition effect may be the result of metallic sulfide formation. The redissolution effect is due to desorption from Fe and Mn or clay minerals and the slow kinetics of the metal-sulfide precipitation, the formation of sulfide complexes, or the formation of organometallic complexes.

7. Under reducing conditions, the release phenomena for P and Si compounds were the same as those in the oxidizing condition, but generally with higher release rates as a result of lower pH under reducing conditions.

8. Under slightly oxidizing conditions, the concentrations of trace metals and nutrients were generally between those of the oxidizing and reducing environments.

9. For long-term experiments, no significant difference was found between the resettled and nonresettled test. The releasing pattern of trace metals and nutrients in the resettled tests was close to that of the same redox conditions in the nonresettled tests. Therefore, the transport phenomena are controlled mainly by the redox condition rather than by the physical characteristics of the sediment.

10. In general, the amount of trace metals released is independent of the gross concentrations of sediments. However, the sediments with higher amounts of nutrients may release higher amounts of nitrogen compounds.

ACKNOWLEDGMENTS

This work was supported by the Army Corps of Engineers, Waterways Experiment Station Contract No. DACW39-74-C-0077. The analytical assistance of Mr. Michael Lee for nutrients and Ms. Lata Bhatt for chlorinated hydrocarbons is greatly appreciated.

REFERENCES

1. MacKenzie, F. T. and R. M. Garrels. "Chemical Mass Balance between Rivers and Oceans," *Amer. J. Sci.* **264**, 507 (1966).
2. Krauskopf, K. B. "Factors Controlling the Concentrations of Thirteen Rare Metals in Seawater," *Geochim. Cosmochim. Acta* **9**, 1-32B (1956).
3. Mortimer, C. H. "The Exchange of Dissolved Substances between Mud and Water in Lakes," *J. Ecol.* **29**, 280 (1941).
4. Mortimer, C. H. "The Exchange of Dissolved Substances between Mud and Water in Lakes," *J. Ecol.* **30**, 147 (1942).
5. Carrit, D. E. "Sorption Reaction and some Ecological Implications," *Deep-Sea Res.* **1**, 224 (1954).
6. Shukla, S. S., J. K. Syers, J. D. H. Williams, D. E. Armstrong and R. F. Harris. "Sorption of Inorganic Phosphate by Lake Sediment," *Soil Sci. Soc. Amer. Proc.* **35**, 244 (1971).
7. Huang, J. and C. Liao. "Adsorption of Pesticides by Clay Minerals," *J. San. Eng. Div., ASCE* **96**, 1057 (1970).
8. Hartung, R. and G. W. Klingler. "Concentration of DDT by Sedimental Polluting Oils," *Environ. Sci. Technol.* **4**, 407 (1970).
9. Boucher, F. R. and G. F. Lee. "Adsorption of Lindane and Dieldrin Pesticides on Unconsolidated Aquifer Sands," *Environ. Sci. Technol.* **6**, 538 (1972).
10. Presley, B. J., R. R. Brooks and I. R. Kaplan. "Manganese and Related Elements in the Interstitial Water of Marine Sediments," *Science* **158** (3803), 906 (1967).
11. Johnson, V., N. Cutshall and C. Osterberg. "Retention of ^{65}Zn by Columbia River Sediment," *Water Res.* **3**(1), 99 (1967).
12. Li, Y. H., J. Bischoff and G. Mathieu. "The Migration of Manganese in the Arctic Basin Sediment," *Earth Planet. Sci. Letters* **7**, 265 (1969).
13. Brooks, R. R., B. J. Presley and I. R. Kaplan. "Trace Elements in the Interstitial Waters of Marine Sediments," *Geochim. Cosmochim. Acta* **32**, 397 (1968).

14. Yemel'yanov, Y. M. and N. B. Vlasenko. "Concentrations of Dissolved Forms of Fe, Mn and Cu in Marine Pore Waters of the Atlantic Basin," *Geochem. Internat.* 9(5), 855 (1972).

15. Presley, B. J. *et al.* "Early Diagenesis in a Reducing Fjord, Saanich Inlet, British Columbia—II. Trace Element Distribution in Interstitial Water and Sediment," *Geochim. Cosmochim. Acta* 36, 1073 (1972).

16. Lee, G. F. and R. H. Plumb. "Literature Review on Research Study for the Development of Dredge Material Disposal Criteria," Contract No. DACW39-74-0024, U.S. Army Waterways Experiment Station, Corps of Engineers, Vicksburg, Mississippi (1974), 130 pp.

17. Wakeman, T. H. "Mobilization of Heavy Metals from Resuspended Sediments," presented at 168th American Chemical Society National Meeting, Atlantic City, New Jersey (September 1974).

18. Waksman, S. A. and M. Hotchkiss. "On the Oxidation of Organic Matter in Marine Sediments by Bacteria," *J. Mar. Res.* 1, 101 (1938).

19. Rittenberg, S. C., K. O. Emery and W. L. Orr. "Regeneration of Nutrients in Sediments of Marine Basins," *Deep Sea Res.* 3, 23 (1955).

20. Kemp, L. W. and A. Mudrochova. "Distribution and Forms of Nitrogen in a Lake Ontario Sediment Core," *Limnol. Oceanog.* 17(6), 855 (1972).

21. Austin, E. R. and G. F. Lee. "Nitrogen Release from Lake Sediments," *J. Water Poll. Cont. Fed.* 45(5), 870 (1973).

22. Zicker, E. L. and K. C. Berger. 'Phosphorus Release from Bog Lake Muds," *Limnol. Oceanog.* 1, 269 (1956).

23. MacPherson, L. B., N. R. Sinclair and F. R. Hayes. "Lake Water and Sediment—III. The Effect of pH on the Partition of Inorganic Phosphate between Water and Oxidized Mud or Ash," *Limnol. Oceanog.* 3, 318 (1958).

24. Pomeroy, L. R., E. E. Amith and C. M. Grant. "The Exchange of Phosphate between Estuarine Water and Sediments," *Limnol. Oceanog.* 10(2), 167 (1965).

25. McKee, G. D. *et al.* "Sediment-Water Nutrient Relationships— Part II," *Water Sew. Works* 117, 246 (1970).

26. Li, W. C. *et al.* "Rate and Extent of Inorganic Phosphate Exchange in Lake Sediments," *Soil Sci. Soc. Amer. Proc.* 36, 279 (1972).

27. Fanning, K. A. and D. R. Schink. "Interaction of Marine Sediments with Dissolved Silica," *Limnol. Oceanog.* 14(1), 59 (1969).

28. Elgawhary, S. M. and W. L. Lindsay. "Solubility of Silica in Soils," *Soil Sci. Soc. Amer. Proc.* 36, 439 (1962).

29. Robertson, D. E. "Role of Contamination in Trace Element Analysis of Sea Water," *Anal. Chem.* 40, 1067 (1968).

30. Rattonetti, A. "Determination of Soluble Cd, Pb, Ag and Zn in Rainwater and Stream Water with the Use of Flameless Atomic Absorption," *Anal. Chem.* 46(6), 739 (1974).

31. Proctor, R. R., Jr. "Stabilization of the Nitrite Content of Sea Water by Freezing," *Limnol. Oceanog.* 7, 479 (1962).

32. Collier, A. W. and K. T. Marvin. "Stabilization of the Phosphate Ratio of Sea Water by Freezing," *U.S. Fish Wildlife Serv. Fish. Bull.* **54**, 71 (1953).

33. Riley, J. P., K. Grasshof and A. Voipio. "Nutrient Chemicals," in *A Guide to Marine Pollution*, E. D. Goldberg, ed. (New York: Gordon and Breach, 1972), pp. 81-110.

34. American Public Health Association. *Standard Methods for the Examination of Water and Wastewater*, 13th ed. (1971).

35. Brooks, R. R., B. J. Presley and I. R. Kaplan. "APDC-MIBK Extraction System for the Determination of Trace Elements in Saline Waters by Atomic Absorption Spectrophotometry," *Talanta* **14**, 809 (1967).

36. Segar, D. A. and J. G. Gonzales. "Evaluation of Atomic Absorption with a Heated Graphite Atomizer for the Direct Determination of Trace Transition Metals in Sea Water," *Anal. Chim. Acta* **58** (1972).

37. Barnard, W. M. and M. J. Fishman. "Evaluation of the Use of the Heated Graphite Atomizer for the Routine Determination of Trace Metals in Water," *Atomic Absorption Newsletter* **12**(5), 118 (1973).

38. Segar, D. A. "The Use of the Heated Graphite Atomizer in Marine Sciences," *Proc. 3rd Internat. Cong. of Atomic Absorption and Atomic Fluorescence Spectrometry* (London: Adam Hilger, 1973), pp. 523-532.

39. Chlorine Institute. "Analytical Methods for the Determination of Total Mercury," Pamphlet No. MIR-104, New York (1970).

40. Jenkins, D. and L. L. Medsker. "Brucine Method for the Determination of Nitrate in Ocean, Estuarine, and Fresh Water," *Anal. Chem.* **36**(3), 610 (1964).

41. American Public Health Association. *Standard Methods for the Examination of Water and Wastewater*, 11th ed. (1965).

42. Armour, J. A. and J. A. Burke. "Method for Separating Polychlorinated Biphenyls from DDT and its Analogs," *J. Assoc. Offic. Anal. Chem.* **53**(4), 761 (1970).

43. Goerlitz, D. F. and W. L. Lamer. *U.S. Geological Survey Water Supply Papers*, No. 1817-C (1967).

44. Mills, P. A. "Variation of Florisil Activity: Simple Method for Measuring Adsorbent Capacity and its Use in Standardizing Florisil Columns," *J. Assoc. Offic. Anal. Chem.* **51**(1), 29 (1968).

45. Reynolds, L. M. "Polychloribiphenyls (PCB's) and their Interference with Pesticide Residue Analysis," *Bull. Environ. Contam. Toxicol.* **4**(3), 128 (1969).

46. Schultzmann, R. L., D. W. Woodham and C. W. Collier. "Removal of Sulfur in Environmental Samples Prior to Gas Chromatographic Analysis for Pesticides Residues," *J. Assoc. Offic. Anal. Chem.* **54**(5), 1117 (1971).

47. Snyder, D. and R. Reinert. "Rapid Separation of Polychlorinated Biphenyls from DDT and its Analogues on Silica Gel," *Bull. Environ. Contam. Toxicol.* **6**(5), 385 (1971).

48. Hubbard, H. L. "Chlorinated Biphenyl and Related Compounds," In *Kirk-Othmer Encyclopedia of Chemical Technology*, Vol. 5, 2nd ed. (1964), pp. 289-297.

49. *Official Method of Analysis of the Association of Analytical Chemists*, 11th ed. (1970), pp. 475-511.

50. Zirino, A. and Y. Yamamoto. "A pH-Dependent Model for the Chemical Speciation of Copper, Zinc, Cadmium and Lead in Seawater," *Limnol. Oceanog.* **17**(5), 661 (1972).

51. Dyrssen, D. *et al.* "Inorganic Chemicals," In *A Guide to Marine Pollution*, E. R. Goldberg, ed. (New York: Gordon and Breach, 1972), pp. 41-58.

52. Morel, F. and J. Morgan. "A Numerical Method for Computing Equilibria in Aqueous Chemical Systems," *Environ. Sci. Technol.* **6**(1), 58 (1972).

53. Ste-Marie, J., A. E. Torma and A. O. Gubeli. "The Stability of Thiocomplexes and Solubility Products of Metal Sulfides—I. Cadmium Sulfides," *Can. J. Chem.* **42**, 662 (1964).

54. Barnes, H. L. and G. K. Czamanske. "Solubilities and Transport of Ore Minerals," In *Geochemistry of Hydrothermal Ore Deposits*, H. L. Barnes, ed. (New York: Holt Rinehart & Winston, 1967).

55. Schwarzenbach, G. and M. Widmer. "The Solubility of Metallic Sulfides—I. Black Mercury Sulfide," *Helv. Chim. Acta* **46**, 2613 (1963).

56. Anderson, G. M. "The Solubility of PbS in H_2S-Water Solutions," *Econ. Geol.* **57**, 809

57. Sillen, L. G. and A. E. Martell. "Stability Constants of Metal-Ion Complexes," *Spec. Pub. No. 17* (1964), Spec. Pub. No. 25 (1971). (London: Chemical Society).

58. Lu, J. C. S. "Studies on the Long-Term Migration and Transformation of Trace Metals in the Polluted Marine Sediment-Seawater System," Ph.D. Dissertation, University of Southern California (1976).

13

ORGANOMETALLIC INTERACTIONS IN RECENT MARINE SEDIMENTS

Miroslav Z. Knezevic and Kenneth Y. Chen

Environmental Engineering Program
University of Southern California
Los Angeles, California 90007

INTRODUCTION

Naturally occurring organic compounds in marine sediments have been the subject of recent interest.[1] Many of these organic substances are known to form complexes with metal ions in solution.[2-5] However, serious analytical problems are encountered in dealing with organometallic interactions in seawater. The sequential chemical dissolution using different chemical agents has been used for this purpose.[6] However, because the hydrogen peroxide treatment leaches both the organic and sulfidic content of marine sediments, the portion of total trace metals associated with each fraction cannot be determined.

A more direct analytical approach involves an actual physical separation of the organometallic complex, by isolating that part of the organic matter which forms the strongest bonds with metals. Humic substances appear to be responsible for these strong interactions.[7-10] Moreover, these substances have been shown to be a major component of the organic matter in recent marine sediments.[11]

Availability of trace metals to biota at the interfaces between seawater and solid sediments is affected by organometallic interactions; the organic substances and seawater may solubilize the insoluble forms of trace metals attached to the solid surfaces through chelation. Organic ligands from biological debris and synthetic organic materials such as carboxyl, carbonyl, amino, amide and mercaptans can bind strongly to the metal ions through coordinate and covalent linkages. Upon incorporation into the sediment, these organometallic complexes may undergo further changes. The release of trace metals due to the disturbance and resuspension of sediment has been the subject of considerable concern, particularly with regard to its relationship to organic matter. Baturin et al.[12] proposed that the alteration of organic matter in anaerobic sediment is the single most important factor in determining the mobility and availability of heavy metals in sediment.

In the case of soils, the migration and accumulation of trace metals has been found to be associated with the presence of humic substances.[7,8,13,14] The different functional groups present in humic substances appear to be responsible for the organometallic complexing phenomenon.[15-17] In this study, efforts were made to determine whether such relationships also exist in near-shore recent marine sediments.

The terms "humic substance" and "humic acid" have been used interchangeably and, perhaps, incorrectly.[18] In this study, a humic substance is considered that portion of the organic matter that contains humic and fulvic acids. The alkalimetric extraction of the humic substances represents the physical separation of the organometallic complexes, as referred to previously. The separation of humic substances into humic acid and fulvic acid is achieved by acidification to pH 1; the humic acid phase is insoluble in acid, whereas fulvic acid is soluble in both acid and base.

MATERIALS AND METHODS

Sediments from seven stations ranging from the sewer outfall to the breakwater in Los Angeles Harbor were extracted for humic substances. The humic substances were extracted according to the procedure outlined by Nissenbaum and Kaplan.[11] Briefly, the sediments were successively extracted with $0.1N$ NaOH, followed by centrifugation at 20,000 g until the supernatant became colorless; then humic-fulvic acid separation was achieved via acidification to pH 1. Some of the extractions required more than one week to complete. In the case of certain organically rich sediments in the Sea of Japan, the extraction of humic substances took weeks, and even months.[19]

The humic acid and fulvic acid were first dried in a rotary evaporator at $35°C$, then in a vacuum dessicator over P_2O_5 at room temperature. Some of the details of sample preparation have been described by Rashid and King.[20]

Infrared spectra of both humic and fulvic acids were taken by the KBr pellet method on a Beckman IR 20-A infrared spectrophotometer. A Perkin-Elmer 305B atomic absorption spectrophotometer, with and without a heated graphite atomizer HGA 2100 (depending upon trace metal concentration), was used to determine all metal concentrations. A deuterium arc background corrector was used in conjunction with the atomic absorption spectrophotometer to remove the effect of nonspecific absorption (due to NaCl) and light scattering in determining trace metal concentrations in the humic substances and humic and fulvic acid fractions. The continuum radiated by the deuterium arc replaces the ordinary reference beam when the arc is turned on. From a large number of extractants studied, dilute aqueous NaOH solution was found to be quantitatively the most effective reagent for extracting large quantities of humic substances from soils or sediments.[21] Since there is some evidence that autoxidation of humic substances may occur under alkaline conditions, oxygen was displaced with nitrogen from the air-tight flask containing the marine sediment.

Organic carbon determinations were performed on the sediment using a LECO TC-12 automatic carbon determinator. Prior to analysis, all samples were treated to remove carbonates.[20] Organic carbon determinations on liquid humic substance, humic acid and fulvic acid samples were performed on a Beckman TOC analyzer equipped with an infrared detector.

X-Ray analysis was performed on powdered fulvic acid using a Phillips-Norelco X-ray diffractometer modified with a curved crystal monochrometer.

RESULTS AND DISCUSSION

Trace metal analyses performed on the humic substances, humic acid and fulvic acid fractions extracted from the Los Angeles Harbor sediments included: Ag, Al, As, Cd, Cr, Cu, Fe, Mn, Ni, Pb, V and Zn. The results are shown in Tables 13.1 through 13.7. Although Stations 1, 3, 4 and 5, and 2 and 7 have the same sediment characteristics, they differ in the state of decomposable organic compounds and in the amount of total organic carbon associated with each sediment (Table 13.8, Column 1).

Iron was found to be in the highest concentration in all sediments, as well as in the humic substances and humic and fulvic acid fractions. Yet the percentage of total Fe concentration in these fractions varies only slightly: between 1.22 and 2.3% for humic substances; between 0.62 and 1.2% for humic acid; and between 0.59 and 1.15% for fulvic acid. Ag and Cd, however, were found to be in the lowest concentration in all sediments and in the humic substance fractions, but were found to be associated with humic substances in larger percentages of the total than was found for Fe. Cd percentages ranged from 0.89 to 4.70% for humic substances; between 0.47 and 2.35% for humic acid; and between 0.32 and 1.66% for fulvic acid. Association of Mn with the humic substance

Table 13.1 Trace Metal Association with Humic Substance, Humic Acid, and Fulvic Acid in a Los Angeles Harbor Sediment (Silty Sand), Station 1

Element	Total Trace Metal Concentration (mg/kg)	Trace Metal with Humic Substance (mg/kg)	Trace Metal with Humic Acid (mg/kg)	Trace Metal with Fulvic Acid (mg/kg)	Percentage of Trace Metals Associated with Humic Substance	Percentage of Trace Metals Associated with Humic Acid	Percentage of Trace Metals Associated with Fulvic Acid
Ag	0.20	0.04	0.02	0.03	20.0	10.0	15.0
Al	11,360	121.1	86.4	32.5	1.06	0.76	0.28
As[a]	6.30	0.86	0.67	0.10	13.6	10.6	1.58
Cd[a]	0.51	0.02	0.01	0.006	3.91	1.95	1.17
Cr	50.6	1.00	0.61	0.32	1.96	1.21	0.63
Cu	21.5	0.82	0.36	0.45	3.81	1.67	2.09
Fe[a]	12,775	215.4	113.6	100.3	1.68	0.89	0.78
Mn[a]	334.1	0.37	0.21	0.13	0.11	0.063	0.039
Ni[a]	12.3	0.45	0.28	0.19	3.65	2.27	1.54
Pb	4.24	0.06	0.03	0.028	1.41	0.71	0.66
V	53.1	1.81	0.83	0.92	3.4	1.56	1.73
Zn	66.6	2.6	1.8	0.70	3.90	2.70	1.05

[a]Trace metal concentration obtained with simultaneous use of background corrector.

Table 13.2   Trace Metal Association with Humic Substance, Humic Acid and Fulvic Acid
in a Los Angeles Harbor Sediment (Sandy Silt), Station 2

Element	Total Trace Metal Concentration (mg/kg)	Trace Metal with Humic Substance (mg/kg)	Trace Metal with Humic Acid (mg/kg)	Trace Metal with Fulvic Acid (mg/kg)	Percentage of Trace Metals Associated with Humic Substance	Percentage of Trace Metals Associated with Humic Acid	Percentage of Trace Metals Associated with Fulvic Acid
Ag	0.66	0.06	0.02	0.03	9.09	3.03	4.54
Al	13,370	93.3	46.5	43.3	0.69	0.34	0.32
As[a]	13.9	1.14	0.90	0.27	8.21	6.47	1.94
Cd[a]	1.77	0.04	0.02	0.015	2.25	1.13	0.85
Cr	121.4	1.80	0.85	0.93	1.48	0.70	0.76
Cu	93.8	1.12	0.83	0.36	1.19	0.88	0.38
Fe[a]	27,450	586.8	263.6	315.8	2.13	0.96	1.15
Mn[a]	455.6	0.26	0.12	0.19	0.06	0.026	0.04
Ni[a]	28.4	0.95	0.63	0.36	3.34	2.22	1.26
Pb	13.1	0.18	0.06	0.06	0.99	0.46	0.46
V	81.9	6.31	4.28	2.23	7.70	5.22	2.72
Zn	180.1	5.10	2.80	2.40	2.83	1.55	1.33

[a]Trace metal concentration obtained with simultaneous use of background corrector.

Table 13.3 Trace Metal Association with Humic Substance, Humic Acid and Fulvic Acid in a Los Angeles Harbor Sediment (Silty Sand), Station 3

Element	Total Trace Metal Concentration (mg/kg)	Trace Metal with Humic Substance (mg/kg)	Trace Metal with Humic Acid (mg/kg)	Trace Metal with Fulvic Acid (mg/kg)	Percentage of Trace Metals Associated with Humic Substance	Percentage of Trace Metals Associated with Humic Acid	Percentage of Trace Metals Associated with Fulvic Acid
Ag	0.283	0.002	0.001	0.0006	0.71	0.35	0.21
Al	14,648	180.3	102.5	70.6	1.23	0.69	0.48
As[a]	9.05	0.38	0.18	0.23	4.19	1.98	2.54
Cd[a]	0.67	0.006	0.0032	0.003	0.89	0.47	0.44
Cr	63.7	0.68	0.32	0.27	1.06	0.50	0.42
Cu	54.2	1.08	0.83	0.42	1.99	1.53	0.77
Fe[a]	19,620	335.6	195.3	156.3	1.71	0.99	0.84
Mn[a]	428.4	0.077	0.05	0.03	0.018	0.012	0.007
Ni[a]	21.4	0.82	0.38	0.45	3.83	1.77	2.10
Pb	7.69	0.63	0.35	0.38	8.20	4.94	4.55
V	69.6	4.87	2.85	2.43	6.99	4.09	3.50
Zn	108.5	5.21	2.86	2.37	4.80	2.63	2.18

[a]Trace metal concentration obtained with simultaneous use of background corrector.

Table 13.4   Trace Metal Association with Humic Substance, Humic Acid and Fulvic Acid in a Los Angeles Harbor Sediment (Silty Sand), Station 4

Element	Total Trace Metal Concentration (mg/kg)	Trace Metal with Humic Substance (mg/kg)	Trace Metal with Humic Acid (mg/kg)	Trace Metal with Fulvic Acid (mg/kg)	Percentage of Trace Metals Associated with Humic Substance	Percentage of Trace Metals Associated with Humic Acid	Percentage of Trace Metals Associated with Fulvic Acid
Ag	0.178	0.003	0.002	0.002	1.68	1.12	1.12
Al	11,910	416.8	278.5	153.6	3.49	2.33	1.28
As[a]	8.21	0.35	0.18	0.19	4.26	2.19	2.31
Cd[a]	0.34	0.016	0.008	0.005	4.70	2.35	1.47
Cr	51.7	0.37	0.26	0.15	0.72	0.50	0.29
Cu	41.7	0.17	0.09	0.06	0.41	0.22	0.14
Fe[a]	20,384	305.7	183.5	124.5	1.49	0.90	0.61
Mn[a]	494.4	0.048	0.033	0.021	0.009	0.006	0.004
Ni[a]	19.5	0.585	0.331	0.263	3.0	1.69	1.34
Pb	7.03	0.30	0.14	0.18	4.26	1.99	2.56
V	50.9	1.78	1.31	0.95	3.49	2.22	1.86
Zn	95.8	4.80	2.8	1.6	5.01	2.92	1.67

[a]Trace metal concentration obtained with simultaneous use of background corrector.

Table 13.5 Trace Metal Association with Humic Substance, Humic Acid and Fulvic Acid in a Los Angeles Harbor Sediment (Silty Sand), Station 5

Element	Total Trace Metal Concentration (mg/kg)	Trace Metal with Humic Substance (mg/kg)	Trace Metal with Humic Acid (mg/kg)	Trace Metal with Fulvic Acid (mg/kg)	Percentage of Trace Metals Associated with Humic Substance	Percentage of Trace Metals Associated with Humic Acid	Percentage of Trace Metals Associated with Fulvic Acid
Ag	0.206	0.002	0.0009	0.0008	0.97	0.43	0.38
Al	22,200	215.4	110.7	108.5	0.97	0.49	0.48
As[a]	13.2	0.53	0.28	0.29	4.01	2.12	2.19
Cd[a]	0.33	0.005	0.003	0.0025	1.51	0.91	0.76
Cr	64.4	1.15	0.83	0.62	1.78	1.28	0.96
Cu	56.1	1.08	0.87	0.38	1.92	1.55	0.67
Fe[a]	31,560	385.4	196.3	186.5	1.22	0.62	0.59
Mn[a]	667.4	0.78	0.54	0.36	0.11	0.081	0.054
Ni[a]	24.2	0.73	0.47	0.41	3.01	1.94	1.70
Pb	13.9	0.15	0.08	0.09	1.07	0.57	0.65
V	84.4	1.10	0.83	0.38	1.30	0.98	0.45
Zn	137.2	5.63	3.62	2.13	4.10	2.64	1.55

[a]Trace metal concentration obtained with simultaneous use of background corrector.

Table 13.6 Trace Metal Association with Humic Substance, Humic Acid and Fulvic Acid in a Los Angeles Harbor Sediment (Silty Clay), Station 6

Element	Total Trace Metal Concentration (mg/kg)	Trace Metal with Humic Substance (mg/kg)	Trace Metal with Humic Acid (mg/kg)	Trace Metal with Fulvic Acid (mg/kg)	Percentage of Trace Metals Associated with Humic Substance	Percentage of Trace Metals Associated with Humic Acid	Percentage of Trace Metals Associated with Fulvic Acid
Ag	—[b]	—	—	—	—	—	—
Al	10,371	105.5	58.5	61.3	1.01	0.56	0.59
As[a]	6.22	0.85	0.53	0.32	13.6	8.5	5.1
Cd[a]	0.18	0.008	0.004	0.003	4.44	2.22	1.66
Cr	58.5	0.45	0.26	0.25	0.77	0.44	0.43
Cu	557.7	9.51	5.41	3.83	1.70	0.97	0.68
Fe[a]	8,034	185.3	96.4	89.5	2.30	1.20	1.11
Mn[a]	336.5	0.41	0.21	0.19	0.12	0.062	0.056
Ni[a]	12.7	0.41	0.23	0.18	3.22	1.65	1.50
Pb	122.3	1.5	0.81	0.63	1.22	0.66	0.52
V	43.3	1.32	0.81	0.36	3.05	1.87	0.83
Zn	212.0	6.81	3.85	2.89	3.21	1.82	1.36

[a]Trace metal concentration obtained with simultaneous use of background corrector.
[b]— not determined.

Table 13.7 Trace Metal Association with Humic Substance, Humic Acid and Fulvic Acid in a Los Angeles Harbor Sediment (Sandy Silt), Station 7

Element	Total Trace Metal Concentration (mg/kg)	Trace Metal with Humic Substance (mg/kg)	Trace Metal with Humic Acid (mg/kg)	Trace Metal with Fulvic Acid (mg/kg)	Percentage of Trace Metals Associated with Humic Substance	Percentage of Trace Metals Associated with Humic Acid	Percentage of Trace Metals Associated with Fulvic Acid
Ag	0.18	0.02	0.01	0.006	11.1	5.55	3.33
Al	21,924	163.3	86.4	78.6	0.75	0.39	0.35
As[a]	5.18	0.27	0.14	0.15	5.21	2.70	2.89
Cd[a]	1.24	0.02	0.01	0.004	1.61	0.81	0.32
Cr	37.1	0.60	0.43	0.32	1.61	1.15	0.86
Cu	54.6	0.96	0.73	0.28	1.75	1.34	0.51
Fe[a]	17,210	346.5	165.5	188.8	2.01	0.96	1.09
Mn[a]	353.5	0.36	0.18	0.15	0.10	0.051	0.042
Ni[a]	17.5	0.68	0.48	0.21	3.88	2.74	1.20
Pb[a]	24.4	0.23	0.13	0.08	0.94	0.53	0.33
V	53.8	1.62	0.85	0.75	3.01	1.58	1.40
Zn	233.6	7.80	5.30	2.80	3.33	2.30	1.20

[a]Trace metal concentration obtained with simultaneous use of background corrector.

Table 13.8 Organic Carbon Content of Sediments, Humic Substance, Humic Acid and Fulvic Acid

Sediments	TOC[a], Total Sediment (%)	TOC, Humic Substance (%)	TOC, Humic Acid (%)	TOC, Fulvic Acid (%)	$\dfrac{TOC_{HS}{}^{b}}{TOC_{sed}{}^{e}}$ (%)	$\dfrac{TOC_{HA}{}^{c}}{TOC_{sed}}$ (%)	$\dfrac{TOC_{FA}{}^{d}}{TOC_{sed}}$ (%)	$\dfrac{TOC_{HA}}{TOC_{HS}}$ (%)	$\dfrac{TOC_{FA}}{TOC_{HS}}$ (%)
Station 1 silty sand	0.379	0.213	0.019	0.125	56.2	5.01	32.9	8.92	58.6
Station 2 sandy silt	0.797	0.389	0.229	0.195	48.8	28.7	24.4	58.8	50.1
Station 3 silty sand	0.497	0.365	0.098	0.222	73.4	19.7	44.6	26.8	60.8
Station 4 silty sand	0.277	0.225	0.060	0.113	81.2	21.6	40.8	26.6	50.2
Station 5 silty sand	0.857	0.335	0.120	0.158	39.0	14.0	18.4	35.8	47.1
Station 6 silty clay	0.343	0.107	0.011	0.039	31.2	3.2	11.4	10.2	36.4
Station 7 sandy silt	1.661	1.800	0.920	0.536	116.1	55.4	32.3	51.1	29.7

[a]TOC = total organic carbon
[b]TOC_{HS} = total organic carbon content of humic substance
[c]TOC_{HA} = total organic carbon content of humic acid
[d]TOC_{FA} = total organic carbon content of fulvic acid
[e]TOC_{sed} = total organic carbon content of sediment

fractions is shown to be low and relatively insignificant in terms of total percentage of Mn. This compares favorably to the findings of Nissenbaum and Swaine.[22] The high concentrations of Al associated with the humic substance fractions may be due to the presence of finely dispersed clayey particles (< 0.1 μm) which were brought into solution during the extraction process.

Among the trace metals studied, As, Cd, Ni, Pb, V and Zn generally showed some significant association with the humic substance fractions, with a few exceptions. Figure 13.1 summarizes the data given in Tables 13.1 through 13.7, as to percentage of total trace metal concentration associated with the various humic substance fractions.

Since wet sediment was used at the start of the successive alkalimetric extraction (containing NaCl in pore water), it was necessary to use the deuterium arc background corrector for elements with wavelengths between 220 and 300 μm. Figure 13.1 shows that both humic and fulvic acids contain all the trace metals. Some loss of trace metals is expected because of the harsh treatment necessary for the physical separation of the different humic substance fractions. This occurs primarily at the point at which humic acid is precipitated out of solution at pH 1 and centrifuged. Therefore, these data should be considered as the low end of the range. However, the bond strengths between the humic acid and metal species might be strong enough to withstand the competition of the hydrogen ion for the replacement of metal ions. The interaction of trace metals with lipids is not included because these molecules lack the ability to interact with trace metals.

Organic carbon content was determined on each sediment, as well as on the humic substance, and humic and fulvic acid fractions. Results are shown in Table 13.8 and Figure 13.2. Generally, trace metal concentrations associated with the humic substances, humic acid and fulvic acid fractions increase as the organic carbon contents remain relatively unchanged (Figure 13.2), with the exception of Station 7, which is located in the proximity of a sewer outfall. Table 13.8, column 5 shows that the total organic carbon in the humic substance accounts for 31.2 to over 100% of the total organic carbon content of the sediment. Humic acid (column 6) ranges from 3.2-55.4% of the total organic carbon; fulvic acid (column 7) ranges from 11.4-44.6%. In general, the fulvic acid fraction constitutes a larger portion of the total organic carbon than does humic acid. This is consistent with the fact that fulvic acid is a precursor of humic acid, so that more fulvic acid is present in recent marine sediments.[23] This same trend is shown in columns 8 and 9, Table 13.8, where it is again seen that fulvic acid generally constitutes the larger fraction of humic substance, with the exceptions of Stations 2 and 7. For Station 2, this phenomenon

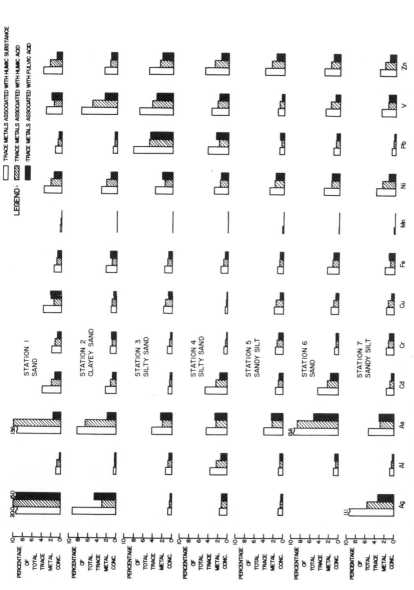

Figure 13.1 Trace metal association with humic substance, humic acid and fulvic acid in seven Los Angeles Harbor sediments.

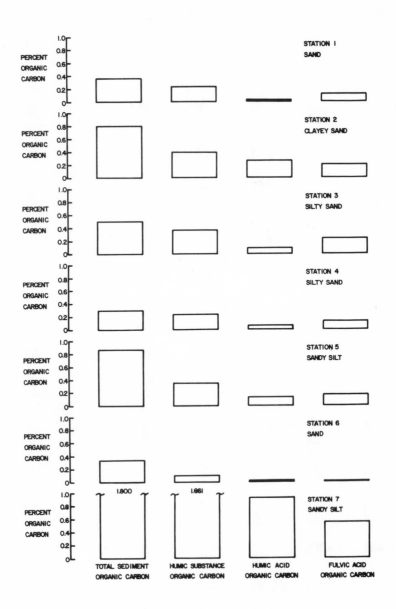

Figure 13.2   Organic carbon contents of the humic substance, humic acid and
fulvic acid with relationship to total organic carbon of seven
Los Angeles Harbor sediments.

is probably due to the presence of clayey particles, which are able to adsorb the fulvic acid and demonstrate the difficulty of the extraction procedure. Since Station 7 is located near an effluent, too many external factors interact on the system to allow a simplified explanation to be proposed.

Differential infrared absorption spectra of the extracted humic and fulvic acids are presented in Tables 13.9 and 13.10. The difference in the absorption bands between extracts from two sediments can probably be explained by the ages of these sediments. For the sediments from Los Angeles Harbor which still contain some decomposable organic compounds, most of the absorption bands are weak or nonexistent toward the lower frequency end of the spectra, while the more stable organic extracts from the midchannel of the San Pedro Basin show significantly different results.

An X-ray analysis was performed on a fulvic acid extract from a silty clay marine sediment. A diffuse X-ray pattern typical of that of noncrystalline substances was obtained. The pattern, as shown in Figure 13.3, consists of a broad maximum around 4.06 Å and a peak at 2.34 Å. Kodama and Schnitzer[24] suggested that 4.06 Å is the distance between sheets of aromatic rings and functional groups such as COOH, OH, CH_3 and C=O. The spacing of 2.34 Å may correspond to another distance between sheets of rings and chains.

Very little definitive information is known at present about the chemical bonds or physical interactions between the trace metals and humic substance molecules such as humic and fulvic acid. Nissenbaum and Swaine[22] proposed that the humic acid molecule is micellar in nature, with an inner hydrophilic cavity holding the trace metal, and outward hydrophobic characteristics. This may be responsible for the impression of enormous bond strengths holding the trace metal in the humic acid molecule. On the other hand, the relative abundance of functional groups, shown in the IR data in Tables 13.9 and 13.10, indicate that these groups are involved in reactions with metals. It is also apparent that humic and fulvic acids can bind metal ions, both through electrostatic forces (attraction of a positively charged metal ion to an ionized COOH group) and by electron pair sharing (formation of a covalent linkage).

In general, the concentration of organic substances that are capable of complexing metal ions in solution is a function of both the physical and chemical characteristics of the water. Lerman and Childs[25] pointed out that the thermodynamic properties of the complexes, as well as chemical exchange between the organometallic complex in solution and any metals held on the surfaces of clays, silts and biogenic material are needed for even a preliminary assessment of organometallic complex phenomena.

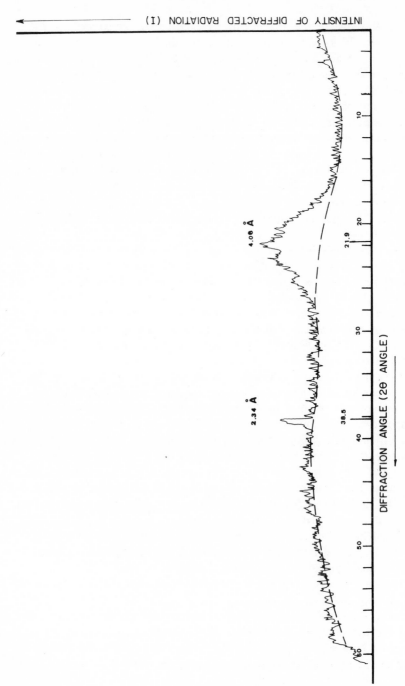

Figure 13.3 X-Ray diffractogram of fulvic acid extracted from a silty clay sediment.

Table 13.9   Infrared Absorption Bands of Fulvic Acid Extracted from Station 6 (Silty Clay) Sediment, Los Angeles Harbor

Absorption Band (cm^{-1})	Structural Features
3400	Hydrogen bonded OH
2900	Aliphatic C-H stretch
1725	C=O of COOH; C=O stretch of ketonic carbonyl
1600	$-C\underset{O}{\overset{O}{<}}$ ionized carboxyl, zwitter ions
1400	COO^{-}
1200	C-O stretch; OH deformation of COOH
800	Unsymmetrical aromatic

Table 13.10   Infrared Absorption Bands of Humic Acid Extracted from Sediments of San Pedro Basin and Los Angeles Harbor (Silty Sand Sediment)

Absorption Bands (cm^{-1})		Structural Features
San Pedro Sediment	Los Angeles Harbor Sediment	
3400	3400	Hydrogen bonded OH
2900	2900 (weak)	Aliphatic C-H stretch
1725 (weak)	1725 (very weak)	C=O of COOH; C=O stretch of ketonic carbonyl
	1630	Aromatic C=C; hydrogen bonded C=O of carbonyl; COO^{-}
1600		$-C\underset{O}{\overset{O}{<}}$ ionized carboxyl, zwitter ions
1200 (weak)	1200 (weak)	C-O stretch
800 (weak)		Unsymmetrical and symmetrical aromatic rings

At present, many questions remain unanswered; for example, the concentration of the metal complex in the solution; the mechanisms and rates of reactions responsible for the degradation of the organic ligands, both in free and complexed form in the solution; the effect of residence time of the water; density stratification; and dispersion within a particular body of water.

Organometallic interactions are obviously complex; extensive work is needed to understand the role of organic substances on the availability and migration of trace metals in marine sediments.

ACKNOWLEDGMENT

This work was supported by the Army Corps of Engineers, Waterways Experiment Station Contract No. DACW39-74-C-0077.

REFERENCES

1. Scheuer, P. J. *Chemistry of Marine Natural Products* (New York: Academic Press, 1973).
2. Degens, E. T. *Geochemistry of Sediments.* (Englewood Cliffs, New Jersey: Prentice-Hall, Inc., 1965).
3. Stumm, W. and J. J. Morgan. *Aquatic Chemistry.* (New York: Wiley-Interscience, 1970), p. 283.
4. Duursma, E. K. "The Dissolved Constituents of Sea Water," *Chemical Oceanography*, vol. 1, J. P. Riley and G. Skirrow, eds. (New York: Academic Press, 1965).
5. Sillen, L. G. and A. E. Martell. "Stability Constants of Metal-Ion Complexes," *Spec. Pub. No. 17* (1964), *Supplement No. 1* (1971) (London: Chemical Society).
6. Presley, B. J., Y. Kolodny, A. Nissenbaum and I. R. Kaplan. *Geochim. Cosmochim. Acta* **36**, 1073 (1972).
7. Drozdova, T. V. *Soviet Soil Sci.* **10**, 1393 (1968).
8. Manskaya, S. M. and T. V. Drozdova. *Geochemistry of Organic Substances.* (New York: Pergamon Press, 1968), p. 354.
9. Schnitzer, M. *Internat. Cong. Soil Sci.* **1**, 635 (1968).
10. Schnitzer, M. In *Organic Compounds in Aquatic Environments*, S. Faust and J. V. Hunter, eds. (New York: Marcel Dekker, 1971), p. 307.
11. Nissenbaum, A. and I. R. Kaplan, eds. *Limnol. Oceanog.* **17**, 570 (1972).
12. Baturin, G. N., A. V. Kockemov and K. M. Shinkus. *Geochem. Internat.* **4**, 19 (1967).
13. Kononova, M. M. *Soil Organic Matter.* (New York: Pergamon Press, 1966), pp. 191-198.
14. Swain, F. M. *Organic Geochemistry I.* (New York: Pergamon Press, 1963), pp. 87-147.
15. Orlov, D. S. and N. L. Yeroshicheva. *Dok. Soil Sci.* **13**, 1799 (1967).

16. Schnitzer, M. and S. I. M. Skinner. *Soil Sci.* **96**, 86 (1963).
17. Schnitzer, M. and S. I. M. Skinner. *Soil Sci.* **99**, 278 (1965).
18. Nissenbaum, A. Personal communication, July 1975.
19. Ishiwatara, R. Personal communication, University of California, Los Angeles, January 1975.
20. Rashid, M. A. and L. H. King. *Geochim. Cosmochim. Acta* **33**, 147 (1969).
21. Schnitzer, M. and S. U. Khan. *Humic Substances in the Environment.* (New York: Marcel Dekker, 1972), p. 9.
22. Nissenbaum, A. and D. J. Swaine. "Organic Matter-Metal Interactions in Recent Sediments: The Role of Humic Substances," submitted for publication, *Geochim. Cosmochim. Acta*, June 1975.
23. Nissenbaum, A. Personal communication, January 1975.
24. Kodama, H. and M. Schnitzer. *Soil Sci. Soc. Amer. Proc.* **31**, 257 (1976).
25. Lerman, A. and C. W. Childs. "Metal-Organic Complexes in Natural Waters: Control of Distribution by Thermodynamic, Kinetic and Physical Factors," In *Trace Metals in Metal-Organic Interactions in Natural Waters*, P. Singer, ed. (Ann Arbor, Michigan: Ann Arbor Science Publishers, 1973), p. 202.

14

ANALYSIS OF ORGANIC CONSTITUENTS OF RECENT AND ANCIENT MARINE SEDIMENTS BY THERMAL CHROMATOGRAPHY

D. A. Scrima, W. C. Meyer and T. F. Yen

Department of Chemical Engineering
University of Southern California
Los Angeles, California   90007

INTRODUCTION

The analysis and characterization of organic components of sediments have usually been restricted to rather laborious and tedious extraction procedures followed by gas chromatography, infrared spectroscopy and mass spectrometry. The use of thermal analysis to study sediments has not been extensively used; pyrolysis-gas chromatography is the most effective method of studying the organics, but as yet ineffective pyrolysis-gas chromatography systems have hampered studies. Specifically, it is rather difficult to reconstruct all the puzzles caused by molecular scission and fragmentation. Flash pyrolysis of the material yields a variety of products and usually results in a radical formation and an uncontrolled set of secondary reactions which yield products that are difficult to relate to actual components.[1,2] In addition, poor reproducibility limits the value of this technique for organic fingerprinting.

As biostationary and other processes in diagenesis begin to condense and polymerize the biological remains to an insoluble material of high molecular weight, studies by conventional methods of chemical analysis become no longer applicable. A rapid and simple method is needed to examine the complex organic constituents by thermal degradation. The coupling of thermogravimetric analysis (TGA) and differential thermal analysis (DTA) with subsequent analysis of the gases evolved by gas chromatography was the first application of programmed heating pyrolysis; TGA-GC was first demonstrated by Chiu[3] in 1968 for the study of polymers and their pyrolysis products. But due to complicated inter-facing of the TGA with the gas chromatograph, small sample size requirements of TGA, condensation of evolved gases in the TGA apparatus, and large dead volumes have impeded the use of this technique in marine sediment analysis.

Thermal chromatography (TC) combines the aspects of programmed thermal degradation (pyrolysis) with gas chromatographic analysis (Figure 14.1) to give a rapid and simple method of thermal analysis of marine sediments.[4] The method is a two-step process of programmed heating

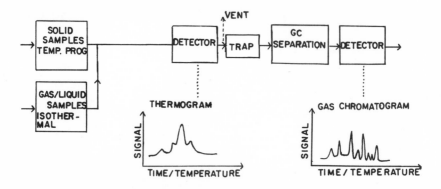

Figure 14.1 Simple schematic of thermal chromatography (TC). The instrument is able to thermally separate a mixture, provide a thermogràm (middle) of each of the resulting products, and finally generate a gas chromatogram (right) of each.

(also flash heating may be done) with monitoring of the gases evolved by either a thermal conductivity detector (TCD) or a flame ionization detector (FID) to provide data on decomposition of the sediment. This information—termed thermogram—yields data comparable to TGA, and to certain aspects of DTA. The information of pyrolysis by TC is equiva-lent to differential thermogravimetric analysis (DTG). As can be seen in Figure 14.2 the top curve is the plot of TGA, the bottom curve is the thermogram derived from TC, and the dashed line is DTG; the dotted line is that of DTA for comparison purposes. The second step of the process is trapping the evolved gas products and subsequently analyzing these products by gas chromatography, thus allowing a two-fold study of a sample of sediment in a matter of a few minutes.

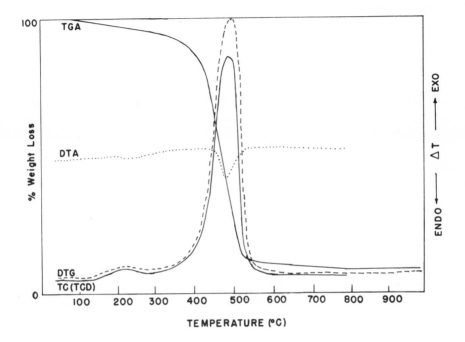

Figure 14.2   Thermal conductivity detector (TCD) used for deriving a thermogram from thermal chromatography (TC) (lower solid line).   Notice the similarity of differential thermogravimetric analysis (DTG) (dashed line).   Also, the transition of differential thermal analysis (DTA) (dotted line) and thermogravimetric analysis (TGA) (upper solid line) are similar.

EXPERIMENTAL

Samples

Two samples of marine sediments were chosen: one from the Santa Barbara Basin, a reducing environment, and the second from the Carmen Basin, an oceanic basin of the Gulf of California.   Several marine sediments of various geologic age were also studied for their diagenic comparison.   The first was a black marine shale from Ohio of Devonian Age from the Olentangy Formation;   another shale of similar age was of deltic origin from the Chattanooga shale;   a sample of oil shale of the Eocene Age from the Green River Formation (Mahogany Zone) was a "nonmarine" sediment but formed from a large inland lake (a fresh water brine deposit). In addition, samples of a number of species of agglutinated foraminifera were analyzed as representatives of chemical fossils, as were various carbohydrate samples.

Apparatus

Thermochromatography was carried out with a model MP-3 Thermal Analyzer/Gas Chromatograph (Chromalytics Corp., Unionville, Pennsylvania). The pyrolyzer (Figure 14.3) is composed of a tubular furnace which

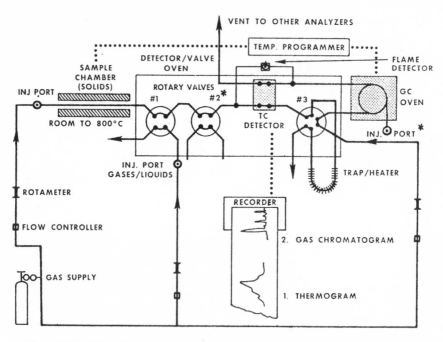

Figure 14.3 Flow diagram of MP-3 of the Chromalytic Corp. The same thermal conductivity detector (TCD) (central) serves both the thermogram and chromatogram. The programably-heated compartment (#2 valve) and port are optional. This is indicated by an asterisk.

allows continuous purging of the chamber with carrier gas (helium) to remove volatile products and prevent further decomposition or reactions to occur. The furnace can be programmed or flash-heated to 800°C and will accommodate up to 1 g of sample for preparative work. From the solids sample chamber, the gas stream is connected to the TCD and FID by valve 1 (Figure 14.3) which allows uninterrupted carrier gas flow to detectors when the solid sample chamber is open to the atmosphere for reloading or when using the liquid injection port. Valve 3 allows for trapping volatile products in the trap for further analysis with the gas chromatograph. The trap is a stainless steel column 1/8-in. o.d. x 16 in., packed with Porpack Q (Waters Associates), a cross-linked polystyrene polymer bead of 80/100 mesh. Trapping may be done at subambient temperatures, if desired, but the trap was operated at ambient temperature. After trapping and pyrolysis are completed, valves 1 and 3 are simultaneously turned to the liquid port and the backflush node with TCD polarity reversed. The trap is then flash-heated to desired temperature (below 250°C) to allow volatiles to be "fired" as an ideal slug into the gas chromatograph.

Along with the gas chromatograph of the MP-3, a model HP 5750 gas chromatograph was also employed for more refined gas chromatograph operating parameters.

PROCEDURE

Pyrolysis Operating Conditions

Approximately 0.050 g of sample was placed in an inert metal boat for the marine sediments and then placed in the fused silica furnace tube, inserted into the solid sample chamber and purged with carrier gas with the furnace closed over the sample for several minutes. The other sediment samples and fossils were placed between two plugs of glass wool in a fused silica tube rather than the boats. TCD current was set at 150 mA and helium carrier gas flow at 20 ml/min. The samples were heated in a programmed manner over the entire range, and some samples were heated only over selected temperature ranges. The evolved gases were trapped at room temperature; with some samples the trapping was done in selected temperature ranges. During the programmed heating, a plot of detector response vs. temperature (or time) for both detectors FID and TCD was generated (the thermogram). Upon completion of pyrolysis, valves 1 and 3 were switched to liquid and backflush and the trap fired to initiate gas chromatographic analysis of the trapped material. In some cases, the gas chromatographic column packing used was not compatible with water. For these cases the trap was first fired with valve 3 in trap position to 105°C to vent the water which was trapped; then valve 3 was turned to backflush position and the trap refired to 240°C.

Chromatograph Operating Conditions

The columns used for gas-solid chromatography are: Porpack Q, 2-ft and 6-ft stainless steel with 1/8-in. diameter; and Chromosorb 101, 4-ft and 6-ft stainless steel with 1/8-in. diameter. For gas-liquid chromatography, 3% SE-30 on Chromosorb Q AW, 6-ft stainless steel with 1/8-in. diameter, and Carbowax 20' M on Chromosorb Q AW, 6-ft stainless steel with 1/8-in. diameter. Both programmed heating and isothermal techniques were employed to achieve optimal chromatographic conditions.

RESULTS AND DISCUSSION

Recent Marine Sediments

Thermograms of two marine sediments are shown in Figures 14.4 and 14.5. They are markedly different in the temperatures at which materials are thermally recovered and the profiles of the products evolved. In the case of the Santa Barbara Basin mud in Figure 14.4 the TCD shows that a larger quantity of water was evolved initially in comparison to the remaining material evolved at a higher temperature. A second small peak at 235°C, detected by TCD but not by FID, indicates inorganic decomposition of some mineral matter. The broad strong peak centered at 495°C signifies the organic decomposition which is evolved, as can be seen clearly by FID. In the case of marine mud from the Carmen Basin (Figure 14.5), the thermogram indicates water evolved at ~ 90°C;

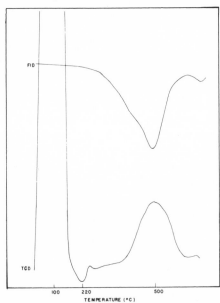

Figure 14.4   Thermogram of marine sediment taken from Santa Barbara Basin. The low curve indicates TCD and the upper curve FID.

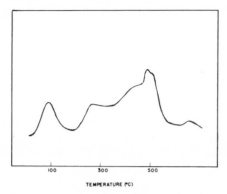

TEMPERATURE (°C)

Figure 14.5   Thermogram of marine sediment taken from Carmen Basin.

however, the intensity of the peak is much weaker and, in comparison with the remaining components, only represents a small quantity of the total evolved material. The other peaks are 270, 495, 515 and 660°C with a shoulder at the rising side of the 495°C peak. Peaks at 495 and 515°C are due to organics.

The evolved material from each sediment was separately trapped, and their chromatograms are shown in Figure 14.6. No attempt was made to identify each peak; rather a comparison was employed for fingerprinting.

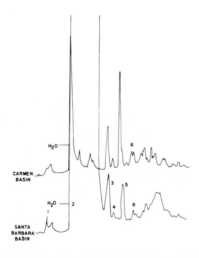

Figure 14.6   Chromatograms of the evolved materials from recent marine sediment. Column of Chromosorb 101 is used for both samples.

There are a number of peaks which are common to both chromatograms: the minor initial peaks prior to the water peak; the water peaks, and peaks 3 through 6. This first indicates the organic material of both basins may be very similar, even though the Santa Barbara Basin is in a reducing environment and the Carmen Basin is in an oxidizing one. Further, in the case of Santa Barbara Basin mud, the broad peak at the extreme right-hand portion of the chromatograph has masked all the peaks which may be similar to those beyond peak 6 of the Carmen Basin sample. Such broadening is commonly caused by an abundance of aromatic components and this pattern may then indicate that organics in the Santa Barbara Basin sample have a higher aromaticity than those from the Carmen Basin. This may be explained by the oxygen-deficient character of the Santa Barbara Basin which would inhibit degradative oxidation of aromatic compounds expected in more normal marine basins.

Ancient Marine Sediments

The thermograms of two Devonian marine shales are shown in Figures 14.7 and 14.8. The peaks at 490°C for both samples are due to organics

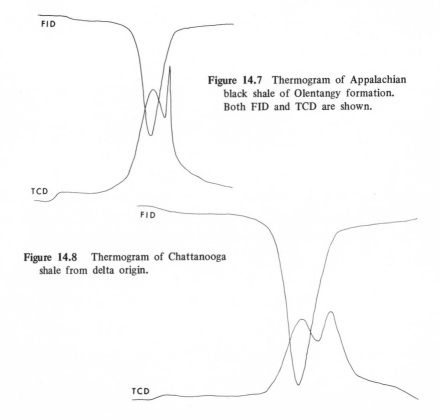

Figure 14.7 Thermogram of Appalachian black shale of Olentangy formation. Both FID and TCD are shown.

Figure 14.8 Thermogram of Chattanooga shale from delta origin.

as indicated by FID in both cases. Furthermore, studies in TGA and differential thermogravimetry (DTG) also supported this assignment.[5] The only difference in the two Devonian shale hinges is the mineral portion as detected by TCD; *i.e.*, the Olentangy shale has a sharp peak at 560°C (Figure 14.7), whereas the Chattanooga shale has a broad peak at ~600°C (Figure 14.8). The 560°C peak is verified as pyrite by addition of pyrite to the sample. The broad peak at 600°C so far remains unidentified. Decomposition temperature of major mineral constituents in shale systems have been previously studied by a number of investigators (Table 14.1). None of those values fall within the range of the broad peak.

Table 14.1   Decomposition of Minerals in Shale System

Temperature, °C	Mineral	Reference
125	Nahcolite	Fisher Scientific Co.[6]
180	Dawsonite	Muller-Vonmoos & Bach[7]
250	Gaylussite	Johnson[8]
250-500	Gaylussite	Johnson[8]
270	Nahcolite	Cameron Eng.[9]
270	Trona	Cameron Eng.[9]
290-330	Dawsonite	Longhman & See[10]
300-375	Dawsonite	Huggins & Green[11]
360-650	Dawsonite	Huggins & Green[11]
	(total decomposition)	
365	Shortite	Johnson, Smith & Robb[12]
370	Dawsonite	Smith & Johnson[13]
370	Dawsonite	Savage & Bailey[14]
375	Gibbsite	Smith[15]
395	Dawsonite	Muller-Vonmoos & Bach[7]
425	Shortite	Johnson, Smith & Robb[12]
470	Shortite	Johnson, Smith & Robb[12]
505	Ferroan gibbsite	Smith[15]
515	Ferroan dolomite	Fisher Scientific Co.[6]
550	Pyrite	Fischer Scientific Co.[6]
595	Fe–ankerite	Smith[15]
640	Mg–ankerite	Smith[15]
670	Dawsonite	Muller-Vonmoos & Bach[7]
700	Calcite	Judd & Pope[16]
750	Ankerite	Fisher Scientific Co.[6]
750	Dolomite	Fischer Scientific Co.[6]
800	Ca–ankerite	Smith[15]
826	Shortite	Johnson, Smith & Robb[12]

The morphology as studied by both petrography and scanning electron microscopy (SEM) suggests that Chattanooga shale differs from Olentangy shale in that the latter contains more well-developed spore bodies.[6,7] A close-up view of the surface of those bodies shows spalls of waxy material intermingled with ground mass. Whether this will contribute to the broadening of the peak observed for the Chattanooga shale is not known at this time.

Similarly, a study of the Green River oil shale indicated that the kerogen decomposition is centered around 480°C (Figure 14.9). A number of mineral peaks, *e.g.*, 270, 390 and 540°C, are present. The latter two peaks are probably due to the presence of dawsonite and pyrite, respectively.

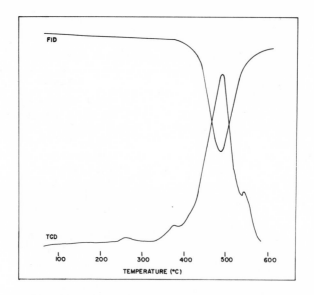

Figure 14.9   Thermogram of Green River oil shale.

The evolved materials trapped have undergone characterization with GC, the resulting chromatograms of which could be used as fingerprinting identifications. For example, a chromatogram of the black shale from the Olentangy formation is shown in Figure 14.10. In contrast, the chromatogram of the oil shale from the Green River formation is illustrated in Figure 14.11.

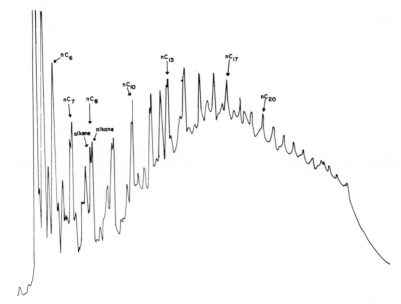

Figure 14.10 Chromatogram of the gas evolved in the thermal decomposition of Appalachian black shale of Olentangy formation (see Figure 14.7 for thermogram).

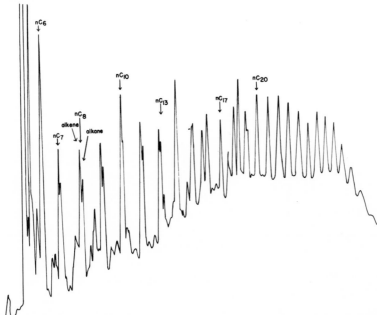

Figure 14.11 Chromatogram of the gas evolved in the thermal decomposition of Green River oil shale (see Figure 14.9 for thermogram).

Benthic Agglutinated Foraminifera

Benthic foraminifera are used by marine scientists as environmental indicators. This group of organisms has been found to be sensitive to environmental parameters such as temperature, salinity, light, depth, oxygen content, nitrogen content and sediment character. In many environs where benthic foraminifera are abundant, immediately adjacent areas with the same physical-chemical character are barren, suggesting that there are other factors affecting distribution within this faunal group. The organics within the sediments may have influence on the foraminifera, either as metabolites, toxins or nutrients. A thorough study of the foraminifera may aid the understanding of the paleoenvironments when organics are deposited.

In order to understand the decomposition of agglutinated foraminifera under elevated temperatures, a few model amino sugars and polysaccharides were chosen for preliminary investigation.

Thermal treatment of standard sugars resulted in degradation, though the fragments were readily separated by chromatography. It becomes apparent that regardless of what standard sugar was chosen, the qualitative yield of breakdown components was the same. These components are, however, released in different quantities, allowing the differentiation of 5- and 6-carbon monosaccharides.

Figure 14.12 compares the thermal chromatographic analysis of glucose (6-carbon monosaccharide) and xylose (5-carbon monosaccharide) on a

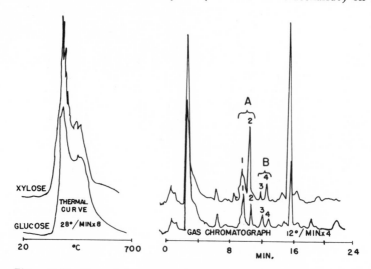

Figure 14.12 Thermal chromatographic analysis of glucose and xylose on a Porapak Q column.

Porapak Q column. The similarities of the pattern are readily visible, but close examination reveals an increase in intensities of several of the peaks for xylose as compared to glucose. In all 5-carbon sugars tested peaks 1 and 2 of doublet A and 3 and 4 of doublet B showed the same size ratios. This relationship remained constant for all the column packings tested. For 6-carbon sugars the pattern of doublets A and B was in all cases reversed.

Sucrose, a dimer of glucose and fructose, was analyzed to ascertain whether it would be possible to resolve sugar monomer units. Initial examination of this chromatograph reveals that the sugars underwent decomposition, yielding results similar to those of pyrolysis[17-19] for that of sucrose. There are five peaks of interest (Figure 14.13), which suggest that the pattern represents a combination of glucose and fructose peaks. Peaks

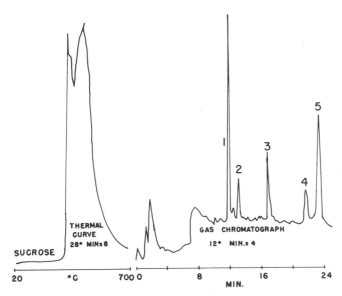

Figure 14.13 Thermal chromatographic analysis of sucrose on Carbowax 20M column.

1, 2, 3 and 5 are emphasized as in fructose, while peak 4 is emphasized as in glucose (Figure 14.14). In addition, the height of fructose in peaks 2 and 5 appears to be minimized in the sucrose pattern, as would be expected by the addition of glucose. Additional experiments are planned utilizing combinations of standards to reproduce the dimer pattern. Chromatographs of foraminiferal polysaccharides revealed a number of components, but were too complex to resolve constituent patterns. Accurate and definite resolution of sugar polymer patterns will, therefore, await computer analysis.

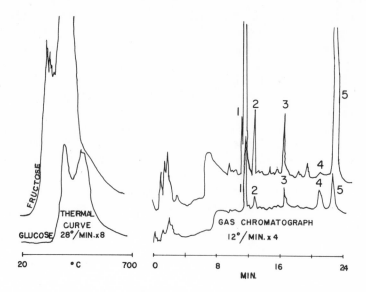

Figure 14.14 Thermal chromatographic analysis of fructose and glucose on Carbowax 20M column.

The thermally liberated components from the organic cement in three species of agglutinated foraminifera are shown in Figure 14.15. Except for minor peaks, *Cribrostomoides, Ammodiscus* and *Hormosina* are similar. On the other hand, the thermograms of *Dentalines, Cribrostomides* and *Ammodiscus* are quite different as shown in Figure 14.16.

CONCLUSION

It has been demonstrated that insoluble organic compounds such as kerogen or the refractory portion of the organic components in marine sediments otherwise amenable to chromatography can, through this technique, be directly analyzed. In this manner it is possible to study recent and ancient sediments in a drilling core from any depth of burial to the surface. The organic components from a formation can also be studied progressively in a vertical direction. The fingerprinting-patterns of the hydrocarbons derived can be used as a tool for prospecting and exploration of fuel sources.

Thermally derived species from sediments can also aid the understanding of the paleoenvironments of the deposition. Many geological markers such as maturation and thermal history can be deduced from the simple tool of thermal chromatography.

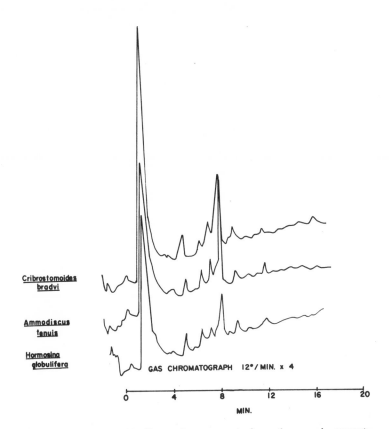

Cribrostomoides
bradvi

Ammodiscus
tenuis

Hormosina
globulifera

GAS CHROMATOGRAPH   12°/MIN. x 4

MIN.

Figure 14.15   Thermally liberated components from the organic cement
in agglutinated foraminifera.

Preliminary studies of the applicability of thermal chromatography in
the analysis of marine sediments reveal that it is possible without chemical separation to isolate and separate numerous sediments associated with
organic compounds and their breakdown products. The feasibility of
differentiating 5- and 6-carbon sugars has been indicated, and it may be
possible to differentiate between different ring conformations of carbohydrate isomers. It is possible that in this manner thermal chromatography
can be used as a tool for dating fossil remains.

Thermal chromatography, when coupled with mass spectrometry or
even mass chromatography, will enable an investigator to identify numerous organic compounds of marine sediments without recourse to classical
procedures presently employed for qualitative chromatographic analysis.

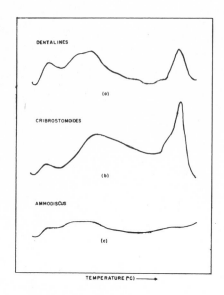

Figure 14.16 Thermograms of the benthic agglutinated foraminifera.

ACKNOWLEDGMENTS

Partial financial support from A.G.A. GR 48-12, NSF GI 35683 and Sea Grant R/RD-2 is appreciated. We also thank Mr. P. L. Warren of Chromalytics Corporation of Unionville, Pennsylvania, for his technical assistance.

REFERENCES

1. Martin, S. B. and R. W. Ramstad. *Anal. Chem.* **33**, 982 (1962).
2. Ettre, K. and P. F. Varadi. *Anal. Chem.* **34**, 752 (1962).
3. Chiu, J. *Anal. Chem.* **40**, 1516 (1968).
4. Scrima, D. A., T. F. Yen and P. L. Warren. *Energy Sources* **1**, 321 (1974).
5. Yen, T. F. Quarterly Report for A.G.A. "Characterization of Oil Shale and Gasification of Bioleached Oil Shale," June 30, 1975.
6. Fischer Scientific Co. "Examination of Fossil Fuels by Thermal Analysis," *Thermofacts*, Bull. TF-31 (1975).
7. Muller-Vonmoos, M. and R. Bach. "Thermoanalytic-Mass Spectrometrical Investigation of an Oil Shale Containing Dawsonite," in *Thermal Analysis*, Vol. 2 (New York: Academic Press, 1969), pp. 1229-1246.
8. Johnson, D. R. "Gaylussite: Thermal Decomposition by Simultaneous Thermal Analysis," *Am. Mineralogist* **58**, 778-784 (1973).

9. Hendrickson, T. A. *Synthetic Fuels Data Handbook* (Cameron Engineering Inc., 1975).
10. Longhman, F. C. and G. T. See. "Dawsonite in the Great Coal Measures at Maswellbrook NSW," *Am. Mineralogist* **52**, 1216-1219 (1967).
11. Huggins, C. W. and T. E. Green. "Thermal Decomposition of Dawsonite," *Am. Mineralogist* **58**, 548-550 (1973).
12. Johnson, D. R., J. W. Smith and W. A. Robb. "Thermal Characteristics of Shortite," USBM Report of Investigation 7862 (1974).
13. Smith, J. W. and D. R. Johnson. *Proc. Second Toronto Symposium on Thermal Analysis*, 95-116 (1967).
14. Savage, J. W. and D. Bailey. "Economic Potential of the New Sodium Minerals Found in the Green River Formation," Symposium on Chemical Engineering Processing, Los Angeles, California (1968).
15. Smith, J. W. Report, ERDA, Laramie Energy Research Center (1973).
16. Judd, M., P. and M. I. Pope. *Thermogravimetric and Gas Evolution Studies of Alkaline Earth Carbonates and Hydroxides*, Vol. 2 (New York: Academic Press, 1969), pp. 1423-1438.
17. Martin, S. B. and R. W. Ramstad. *Anal. Chem.* **33**, 982 (1961).
18. Tai, H., R. M. Powers and T. F. Protzman. *Anal. Chem.* **36**, 108 (1964).
19. Groten, B. *Anal. Chem.* **36**, 1206 (1964).